Lecture Notes in Networks and Systems   **663**

Series Editor

Janusz Kacprzyk, *Systems Research Institute, Polish Academy of Sciences, Warsaw, Poland*

Advisory Editors

Fernando Gomide, *Department of Computer Engineering and Automation—DCA, School of Electrical and Computer Engineering—FEEC, University of Campinas—UNICAMP, São Paulo, Brazil*
Okyay Kaynak, *Department of Electrical and Electronic Engineering, Bogazici University, Istanbul, Türkiye*
Derong Liu, *Department of Electrical and Computer Engineering, University of Illinois at Chicago, Chicago, USA*
   Institute of Automation, Chinese Academy of Sciences, Beijing, China
Witold Pedrycz, *Department of Electrical and Computer Engineering, University of Alberta, Alberta, Canada*
   Systems Research Institute, Polish Academy of Sciences, Warsaw, Poland
Marios M. Polycarpou, *Department of Electrical and Computer Engineering, KIOS Research Center for Intelligent Systems and Networks, University of Cyprus, Nicosia, Cyprus*
Imre J. Rudas, *Óbuda University, Budapest, Hungary*
Jun Wang, *Department of Computer Science, City University of Hong Kong, Kowloon, Hong Kong*

The series "Lecture Notes in Networks and Systems" publishes the latest developments in Networks and Systems—quickly, informally and with high quality. Original research reported in proceedings and post-proceedings represents the core of LNNS.

Volumes published in LNNS embrace all aspects and subfields of, as well as new challenges in, Networks and Systems.

The series contains proceedings and edited volumes in systems and networks, spanning the areas of Cyber-Physical Systems, Autonomous Systems, Sensor Networks, Control Systems, Energy Systems, Automotive Systems, Biological Systems, Vehicular Networking and Connected Vehicles, Aerospace Systems, Automation, Manufacturing, Smart Grids, Nonlinear Systems, Power Systems, Robotics, Social Systems, Economic Systems and other. Of particular value to both the contributors and the readership are the short publication timeframe and the world-wide distribution and exposure which enable both a wide and rapid dissemination of research output.

The series covers the theory, applications, and perspectives on the state of the art and future developments relevant to systems and networks, decision making, control, complex processes and related areas, as embedded in the fields of interdisciplinary and applied sciences, engineering, computer science, physics, economics, social, and life sciences, as well as the paradigms and methodologies behind them.

Indexed by SCOPUS, INSPEC, WTI Frankfurt eG, zbMATH, SCImago.

All books published in the series are submitted for consideration in Web of Science.

For proposals from Asia please contact Aninda Bose (aninda.bose@springer.com).

Artem Antonyuk · Nikita Basov
Editors

Networks in the Global World VI

Proceedings of NetGloW 2022

Editors

Artem Antonyuk
Centre for German and European Studies
Bielefeld University
Bielefeld, Germany

Nikita Basov
Department of Social Statistics, School
of Social Sciences
University of Manchester
Manchester, UK

ISSN 2367-3370   ISSN 2367-3389 (electronic)
Lecture Notes in Networks and Systems
ISBN 978-3-031-29407-5 ISBN 978-3-031-29408-2 (eBook)
https://doi.org/10.1007/978-3-031-29408-2

This Springer imprint is published by the registered company Springer Nature Switzerland AG
The registered company address is: Gewerbestrasse 11, 6330 Cham, Switzerland

Preface

During the past ten years, 'Networks in the Global World' (NetGloW) conference series was organized by the Centre for German and European Studies at St. Petersburg State University and Bielefeld University. The conference brought together network researchers across disciplines and subject areas, enabling analyses of global processes as well as theoretical and methodological advancements.

The conference series took off in 2012, with the meeting subtitled 'Structural Transformations in Europe, the US, and Russia'. It joined researchers, policymakers, and businesspeople to reflect on the changes induced by the growing importance of networks worldwide. NetGloW 2014 focused on bridging theory and methods in network analysis. In 2016, the conference topic was concerned with relations between different types of networks. Two years later, we came together to discuss the principles, logics, and mechanisms generating network structures. NetGloW participants of 2020 discussed the contexts in which networks are embedded and how these contexts constitute both structure and meaning of networks.

The principal theme of the sixth NetGloW edition held on June 22–24, 2022, was time, would it imply change of network structures over time, certain historical points when networks occur, relational stories endowing links and networks with meaning, or other related issues. In line with its tradition, the conference engaged dynamics of different kinds of networks, whether they connect persons, symbolic elements, material things, organizations, social institutions, political entities, or other types of nodes. The event also sought an understanding of relational dynamics between networks of different kinds. We were equally curious to look into network change throughout the development of personal trajectories, relational dynamics in dyads, or at the whole-network level. We were interested to understand the meaning of change in the context of a particular historical time and to look for the fundamental principles of change in typical relations, such as friendship or power.

The participants were asked to propose new theorizations of time in network analysis, offer new methods to test theories of network change, or come up with applications to new data. Archive-based data on networks of the past, longitudinal, and time-stamped data were particularly welcome, especially those revolving around European societies. In addition, the conference not only questioned how network structure changes, but also how individual, dyadic, and whole-level perceptions of specific relations and whole networks change over time, alongside the change of network structure. Accordingly, both the methods capturing change interpretively and those relying on formal and statistical inquiries were welcome, especially mixed methods striving at finding the fundamental principles of network dynamics together with the corresponding network patterns while understanding the meaning of these principles and patterns in broader socio-cultural contexts.

The highlight idea of the conference series remained the same: Combining diverse methods and data to study various types of networks across cultures, societies, states, economics, and cities—with a primary focus on European societies. Traditionally for NetGloW, a particular emphasis was on interconnecting theory, method, and applications, testing theory-driven principles with the help of reflexively chosen methods and datasets.

This volume is proud to deliver a rigorous selection of papers presented at 'Networks in the Global World 2022' and covering five thematic areas: culture and discourse networks, spatially embedded networks, online networks, political and power networks, and professional networks. The papers contribute to the focal topic—networks and their contexts—as well as to the methodology of network analysis and its applications across disciplines and subject areas.

We would like to thank all the authors for presenting at NetGloW'22 and for preparing chapters for this book, despite the challenges of today. We express our gratitude to dozens of reviewers who contributed to delivering the volume. We rest assured that the volume will further inspire network scholarship across cultural, disciplinary, and organizational boundaries in the—so visibly global and networked—world.

Nikita Basov
Artem Antonyuk

Organization

Programme Committee

Nikita Basov (Chair)	University of Manchester, UK
Artem Antonyuk	Centre for German and European Studies, Bielefeld University, Germany
Elisa Bellotti	University of Manchester, UK
Svetlana Bodrunova	St. Petersburg State University, Russia
Iina Hellsten	University of Amsterdam, Netherlands
Olessia Koltsova	HSE University – St. Petersburg, Russia
Oleksandra Nenko	University of Turku, Finland
Camille Roth	CNRS/Humboldt University Berlin, Germany
Peng Wang	Swinburne University of Technology, Australia

Reviewers

Alieva, Deniza	Webster University Tashkent, Uzbekistan
Ananin, Denis	Moscow City University, Russia
Antonyuk, Artem	Centre for German and European Studies, Bielefeld University, Germany
Basov, Nikita	University of Manchester, UK
Bodrunova, Svetlana	St. Petersburg State University, Russia
Gradoselskaya, Galina	National Research University Higher School of Economics, Russia
Hellsten, Iina	University of Amsterdam, Netherlands
Judina, Darja	St. Petersburg State University, Russia
Maltseva, Daria	National Research University Higher School of Economics, Russia
Mayerhoffer, Daniel	Otto-Friedrich-University Bamberg, Germany
Nenko, Oleksandra	University of Turku, Finland
Nigmatullina, Kamilla	St. Petersburg State University, Russia
Ortiz, Francisca	Millennium Institute for Care Research (MICARE), Chile
Pezoldt, Kerstin	Technische Universität Ilmenau, Germany
Poluboyarinova, Larisa	St. Petersburg State University, Russia
Ryabchenko, Natalia	Kuban State University, Russia
Schulz, Jan	Otto-Friedrich-University of Bamberg, Germany

| Sinitsyn, Nikita | Lomonosov Moscow State University, Russia |
| Smoliarova, Anna | St. Petersburg State University, Russia |

Organizer

Partners

Contents

Culture and Discourse Networks

Discourse of Complaining on Social Networks in Russia: Cumulative Opinions vs. Decentering of Institutions

Kamilla Nigmatullina(✉) ⓘ, Svetlana S. Bodrunova ⓘ, Nikolay Rodossky ⓘ, and Dmitry Nepiyushchikh

St. Petersburg State University, St. Petersburg, Russia
k.nigmatulina@spbu.ru

Abstract. Social networks have become a platform for expressing dissatisfaction, support, and social tensions in general. During the pandemic of COVID-19, the audiences' need to find solutions and answers has put heavy burden on authorities and professional journalists. The study addresses the question of to what extent a social network can provide space for deliberation in tackling social issues that organizes the public dialogue for problem solving. Also, we ask whether traditional media and political actors preserve their important roles as major deliberative actors. For answering these questions, we have conducted three-step research. On the first stage, we qualitatively assessed the complaints and responses to them in media-like accounts on VK.com and Instagram, local media, and official portals, as well as conducted 21 structured interviews to contextualize the practice on online complaining in Russia. Then, we collected user comments to posts that contained complaints from 63 accounts on VK.com in 21 regions of November 2020 and February 2021. Via textual analysis, we defined the dominant topics of complaints and the dominant discourse around complaints, as well as the potential for growth of conflict or possible harmonization of discussions. By expert opinions, local media and authorities react differently to the increase in the intensity of complaints. They feel pressure from the platform audiences to increase their involvement. Despite this, neither the nature of the discussions nor the roles of media and authorities' accounts help turn the discussions into deliberative spaces. We have discovered an institutional vacuum in the VK.com discussions, as well as nearly complete absence of deliberative discussion patterns. More often, user comments produce cumulative opinion spaces within complaint-containing commenting, quite in opposition to the normative view of deliberation processes on social media. The result of smoothing out emotions is a fragmented, even if intense, discourse where solutions are not discussed.

Keywords: Social networks · Complaints · Deliberation · Cumulative deliberation · Decentering of journalism · Local journalism · VK.com · Social network analysis

1 Introduction

Social networks in Russia have become not only a source of hyperlocal news in Russian regions but also a space to share emotions and, in particular, complaints. Users utilize accounts of officials and politicians to express their discontent. This, allegedly, leads to changing the chain of aggregation and articulation of public claims [1] from 'people – media – government' to 'people – government – media', as local media, including pro-governmental ones, tend to use the officials' social media accounts as a source of human-interest stories based on complaints.

Today, hyperlocal media, including local newsgroups and similar accounts on social media, play a huge role in news flows [2], but there is no answer to what extent they invest in democratization in countries with no long democratic tradition. Speaking about Russian hyperlocal local newsgroups as 'the new entrants in the local media space of the Russian province that have recently become important actors of regional public communication' [3: 1] on social networks, we should underline that their potential for democratic discussions is challenged by local authorities' information policies and activities. That is, the platforms intended to be open for discussions, in fact, see their audiences to be limited in expression by many factors, from platforms' rules and affordances to legislation punishing the spread of fake news.

The purpose of this study is to explore the role of social networking platforms in the formation of engaged deliberative communities of local citizens, as well as the role of media and authorities' accounts in organizing substantial discussions around user complaints. To do this, we describe how journalists use social networks in the work of hyperlocal media, and what role criticism and complaints disseminated in social networks play in the local news discourses. Then, we formulate exploratory research questions on responses of authorities and media to user complaints, the topicality of complaints, and the perceived roles of media and authorities in dealing with the users' discontent.

To address them, we analyze 180 posts collected from 30 local newsgroups on VK.com (ex-VKontakte) and Instagram most popular in regions of Russia, as well as comments to them (15,299 altogether) and 21 structured interviews with regional experts and officials. The chosen accounts were especially popular in the selected regions. Of these, we selected posts that gathered the most massive user response. Descriptive statistics and interpretive reading techniques were used to receive the answers to the research questions. Network analysis was applied to show how discourses of complaints are being shaped.

The regions have been chosen according to Natalia Zubarevich's concept of 'four Russias' which represent different clusters of industrial and social development [4]. Selected regions reflect diversity of media markets and landscapes, as well as of audiences' demands of critical discussions in social networks. To finalize the list of regions, we have used rankings of social media accounts by Medialogiya and BrandAnalytics media analysis agencies: Arkhangelsk, Dagestan, Irkutsk, Kaliningrad, Karelia, Krasnodar, Krasnoyarsk, Kursk, Leningrad region, Moscow region, Murmansk, Nizhny Novgorod, Northern Ossetia, Novosibirsk, Rostov-on-Don, Ryazan, Samara, Surgut, Tyumen, Velikiy Novgorod, Yakutsk.

By the discourse of complaining, we mean a set of user comments published under news messages on public pages within social networks, aimed at drawing the attention of the audience, journalists, and local authorities to social problems in a particular region. The discourse of complaining is based on two key intentions: First, to share emotions and get support from the audience of the newsgroup, and second, to get the attention of significant communication actors, raise public awareness, and to achieve positive changes in real practice. However, according to the experts (including those from BrandAnalytics), a leader in social media analytics in Russia, only a very limited number of users return to their complaints on social networks, even if the latter receive a response from decision makers. Moreover, only few complaints have a potential to affect policymaking on higher levels, as mostly they remain unheard outside of social networks. This is why the factors that affect the efficiency of user complaints must be researched upon, and, among them, the activities of traditional deliberative actors within the new communicative realms are of special importance.

2 Complaining as a Media Practice

2.1 Deliberation or Opinion Accumulation? The Community-Building Capacity of Social Networks

In Russia, the audiences of hyperlocal social media, as a rule, do not form clearly shaped communities. These are scattered individuals who use local platforms to take out discontent and criticism, more often for the purpose of emotional relaxation than for real problem-solving. High numbers of commenting on social media groups may be mistaken for user engagement; however, despite the popularity and virality of some single posts, we need to ask whether there is true engagement of local audiences in discussing the issues relevant to them on hyperlocal news sources (both the hyperlocal media of traditional provenance and the newsgroups on social media), which would form (hyper)local deliberative communities. As the concept of cumulative deliberation suggests, opinions online form mostly via gradual accumulation and aggregation of homophilic views (if at all detected within white noise), rather than via substance-oriented, rational, and consensus-aiming round-robin deliberation [5]. Thus, exploring the nature of discussions around complaints which represent the public sore points may cast light on how deliberation works on the micro-level; in particular, whether the discussion patterns are traditionally deliberative or cumulative.

As earlier theory suggests, the community-bound relations between (hyper)local media of various nature, their audiences, and the local decision-makers reshape the agenda-setting process in local communities, making this process a participatory one where 'members of the general public can... act as advocates within their respective direct environments, by advancing the agenda in question and playing an active part in the relevant transformation processes' [6]. In the online realm, citizens can do it via more sophisticated forms like e-petitioning [7] or socially mediated protest actions with policy demands [8], or via simpler individual contributions like posing questions and complaints directly to authorities, thus bypassing media as traditional aggregators and articulators of dissent (see below) [1]. At the same time, user complaints may provide input for media agendas which, in their turn, influence decision-making.

Social media have also altered the classic agenda-setting by opening the gate for all possible motives and messages, including both verified and fictional. Unlike before, news streams are filtered not by professional communicators but by secondary and tertiary gatekeeping actors [9] including collective intelligence and algorithms based on user preferences. As a result, local news media find themselves on the periphery of viral content, consumed by rare enough people interested in having a professional filter and classic news selection compared to algorithmic feeds. Such people subscribe to groups and public pages directly to receive updates (as the newsfeed is no longer the most popular way to interact with news-like content). The role of local media, therefore, is not about giving voices to the voiceless but about providing updates on issues of current public concern. As a result, the media space of small towns is an inextricable combination of professional and amateur media, where a media that reacts more actively to the audience feedback becomes influential.

2.2 Decentering of Traditional Deliberative Actors on Social Media

Socially mediated complaints are also a peculiar object of study today. They are often analysed in the context of service science, management, and engineering to help companies receive customer feedback faster, respond to it, and improve the quality of their products and services. Researchers have suggested that 'social media teams should be given continuous and mandatory opportunities to learn to provide high-quality complaint resolutions faster' [10: 337]. Also, scholars have stated that social media is a 'relevant complaint channel', pointing out that 'more than half of the complainants chose to voice their dissatisfaction first via social networks' [11: 534].

At the same time, other scholars have pointed out to an objective process of 'decentering of journalism' in the public sphere and its substitution by audience-led public deliberation [12, 13]. Being an objective trend, decentering of journalism may still have, to their viewpoint, its substantial hazards and shortcomings, as actors legitimized by centuries of professional development (including media with their professional ethics) are moved to discursive and gatekeeping peripheries by collectives with doubtful legitimacy and no shared rules of the game, open to radicalism and irresponsibility. This clearly poses two questions which we focus upon also (but not exclusively) in this paper. First, it is the question of audiences' deliberative quality; second, it is that of the role of traditional deliberative actors, mostly media and authorities, within the new audience-centered deliberative milieus.

Among other papers (see our review in [14]), our earlier research on Twitter shows that media accounts on social networks play important roles in the discussion structure. We have identified two main roles of media and journalists' accounts, namely informing and opinion conveying. Another structural role of media accounts was linking the centers of the assessed discussions to their respective peripheries. We have shown that these functions showed up independent of national context, in all the ad-hoc discussions on conflictual events.

Unlike media accounts, political actors' online presence and their functions within discussions vary significantly in different socio-political contexts. Thus, we showed that, in Russia of the 2010s, there was an institutional vacuum on Twitter, and it differed from other contexts like Germany [15]. The patterns of public blaming and assignment of responsibility for ethnic tensions also differed: Blame was put to the national political actors, but responsibility was expected from regional authorities, while in Germany it was vice versa [16]. All this research has been done for Twitter, and it is important to know whether media preserve their crucial roles also on other platforms like VK.com, and whether political actors are equally absent from online discussions on this social network.

2.3 Complaining in (Semi-)authoritarian States: Transgression of Dissent and Co-optative 'Gardening' of Active Audiences

Some (semi-)authoritarian regimes, the researchers have written, are quite sensitive to the complaints of their residents. For example, it is known that the Chinese state attempts to listen to the complaints of residents against representatives of local authorities, while not giving residents too much power over officials. China sees this as a guarantee of the stability of the regime [17]. The national authorities successfully utilize the co-optation strategy by executing power over local ones based on local communities' complaints, including those online [18], thus making dissent in local communities transgress to the national level but serve for regime consolidation and building trust to it. It has also been shown that, in the states with autocratic trends, comment sections 'may serve political elites for the purposes of gathering information about society, credibly increasing the transparency of government, monitoring lower-level officials or showcasing widespread regime support' [19: 15]. In several countries of the post-Soviet space, the state may easily close one or another media resource on which many critical comments are posted, but practice shows that the authorities rarely resort to this.

However, the quality of criticism in the post-Soviet countries also needs to be questioned. Thus, in the example of Belarus, Bodrunova and colleagues [20, 21] have shown that criticism towards leadership rarely implies substantial, rational, and constructive policy criticism. Complaints about particular problems and issues, thus, may fill in the gap in feedback created by absence of constructive policy-critical publics combined with excessively emotional and insubstantial criticism towards leadership in leadership-critical publics [22].

Counter-intuitively, from an elite's perspective, the benefits of critical comments outweigh the risks. How can critical comments be useful to the authorities? First, on aggregate, they represent not only the opinion of the opposition but also the voice of the entire society. Second, changes in online opinions can be tracked in real time. Third, comments are saved, so changes can always be tracked backward over time. Fourth, voicing criticism in comments may work for 'letting the steam off the system' as an alternative to street protests. In general, 'critically commenting publics can thus be assumed to be associated not only with risks but also with benefits for authoritarian leaders' [23: 489–491].

Nevertheless, much depends on the specific national context. Litvinenko and Toepfl [24] have identified two main strategies for the behavior of a (semi-)autocratic regime in relation to critical user commenting on news media websites, which are repression and integration. Empirical studies have shown that, for example, Azerbaijan has for long been fully focused on the first strategy, trying to minimize audience participation in commenting on news. In turn, Belarus, before 2020, demonstrated adherence to the integration strategy. In Belarus, 'comment sections on some leading news websites, such as, for instance, tut.by, developed into a vibrant space for citizen participation, where users could criticize – within certain boundaries – specific policies, lower-level officials, and even the authoritarian leadership as such' [24: 17–18]. This situation changed after the summer of 2020; in 2021, major news portals around the country were forced to close.

In Russia, Runet is still perceived as a space freer than traditional media in terms of expression of opinions. Until the early 2010s, it enjoyed scant regulation [25] and was an arena where alternative-agenda media and political blogging flourished [26]. Since 2014, a range of restrictive laws have been introduced that have reshaped the communicative climate of Runet [27]. Nonetheless, due to a high share of the state in both federal and regional media markets, professional news media are often perceived as state-affiliated (and thus subjected to the authorities), which prevents them from being seen as platforms suitable for complaints and criticism. In their turn, non-state-affiliated local newsgroups attract active audiences who come to them not in search of information but in search of involved counterparts for discussion.

The political system in Russia has not yet elaborated a unified strategy for public reaction to complaints, especially in risky cases of negative reactions to comments by members of local authorities. Thus, in Krasnoyarsk (Siberia), after an advisor to city government posted on Facebook his highly negative reaction to users' complaints to the wildfires of summer 2021, the local administration denounced the post as his personal, not administrative, position. Such two-faced behavior of both co-opting comfortable enough criticism and denouncing strong reactions from members of government corresponds to the 'gardening' nature of the Russian (semi-)authoritarianism [28]. This is why exploring the roles of local authorities in the discussions around user complaints may help shed light on how such strategies are used and whether they have any impact online.

Media, in their turn, also tend to respond to user complaints – either by creating publications directly based on user-generated content or by taking users as sources. In our research, we also examine media responses to user complaints and see whether media play important roles in the discussions enacted by complaints.

2.4   Previous Studies by the Authors on Social Networks in Russia

Previous research by the authors [29] shows that public communication on local politics is moving into social networks and is constantly monitored by local authorities. We have stated the controversial nature of social networking sites in deliberative terms. On one hand, before 2022, social networks were a place for relatively free expression of critical views and represented some critical publics. On the other, on social media, the emotional degree of criticism was reduced and did not spill over offline, making social networks a tool for venting dissent.

However, the degree of criticism varied throughout the country. Thus, we have detected 'three critical Russias': namely, the regions that freely criticized political decisions ('policy-critical'), regions that freely criticized political leaders ('leadership-critical'), and regions without a request for criticism ('uncritical'), which clearly corresponds to Toepfl's theory of three types of authoritarian publics [22]. Our work contributes to his theory of publics, as we have discovered the spatial (regional) dimension in formation of critical publics. Moreover, this was true not only for the 'native Russian' social network VK.com; by 2021, Instagram had also become a source of official information for both citizens and journalists in many regions.

Another important conclusion was that Russia was extremely heterogeneous in its citizens' demand for complaint and criticism opportunities. Complaints as content are typical for large and economically active regions where there is a clash of political and economic interests, while neutrality and a demand for non-confrontational dialogue prevail in the rest of the territories.

The next stage of our study [30] has demonstrated that social media content is highly flexible in terms of combining emotions and elements of rational argumentation. When it comes to scenarios of the future, rather than to criticizing the present, social media users find themselves more involved in commenting on content, especially if it relates to their fears or false hopes.

Local news media on digital platforms are focused on the main requests of the mass audience for operational information, social navigation and orientation, mobilization, and integration, facilitating communication between the authorities and society, in-forming about the latest events and the situation with COVID-19. The COVID-related scenarios of the future clearly demonstrated the audience's orientation towards emotional rather than rational argumentation of their positions.

Investigating how Russian citizens in the regions expressed their imagined scenarios for the possible future development of the pandemic on social networks, we came to several important conclusions. In particular, we have seen that social media were primarily used for sharing emotions but not information on current events. The key emotions were fear and hope. Complaints formed the basis of many scenarios of future that came into our field of vision. We have come to the conclusion that the complaints expressed on social networks have high potential for provoking and whipping up the corresponding emotions. These emotions, in turn, engage the audience in the discussion.

2.5   Research Questions and Hypotheses

As our study is exploratory, we have posed research questions and hypotheses that can be answered in a mixed-method way. They include the following:

RQ1. How is the discourse of complaints shaped on social networks?
H1.1. Comment conglomerates do not form deliberative communities in most cases. The dominant pattern of commenting is cumulative (represented by conglomerates of non-interlinked comments), not deliberative.
H1.2. The deliberative pattern of discussing is lost with the growth of the number of commenters and comments in the discussions.
H1.3. The type of media (legacy media vs. newsgroups) affects the discussion structure; media accounts differ from newsgroups in how the discussion is shaped.
RQ2. What is the role of media and political accounts in the discussions around complaints?
H2.1. Presence of media and political accounts in the accounts with cumulative commenting differs from that in the accounts with deliberative commenting. It also differs depending on the media type (legacy media vs. newsgroups).

H2.2. In the accounts with deliberative commenting, the role of media and political accounts as commenters is the key one, while, in the cumulative ones, it does not shape the dialogue.

H2.3. The experts' views and the results of social network analysis do not correspond to each other in assessment of the roles which professional media play in the discourse of complaining.

RQ3. How is the institutional response to complaints organized? In case if the media and authorities do not participate much in online discussions, do they respond to user complaints in other ways?

3   Methods and Sampling

3.1   Sample: Selection of Accounts

VKontakte (nowadays shortened to VK.com) is the most popular social network in Russia, whose multi-channel monthly audience fluctuated around 40 mln in 2022–2021 [31]. VK.com is widely used in Russia, with its' role after February 2022 increasing rapidly and significantly. Moreover, it is VK.com that has been the digital home for many hybrid local media [29], as well as for local discussion groups and newsgroups [3]. That is why this social network is the primary one for discussing the role of local news media and social networks in how the culture of complaining becomes a part of the deliberative culture in Russia.

Instagram 'has significantly transformed over the past few years and begun to play a noticeably larger role in the formation of online publics', including in Russia [32: 2]. In the recent years, local Russian authorities have become interested in the use of this social network, so Instagram, for some time before 2022, became a source of official information for both citizens and journalists in many regions. This conclusion was valid until March 2022 when the social network was officially banned in Russia.

To assess the discourse of complaining, we have used the following sampling strategy. There were two datasets of comments from selected accounts, the preliminary one and the main one. First, we have monitored both social networks to define which platform – VK.com or Instagram – hosted the most popular local newsgroups in each region, as platform preference has formed non-systemically region by region. Second, we have chosen 30 most user-engaging media-like accounts, either on VK or on Instagram, in 10 Russian regions (that is, three accounts per region). In each region, with the help of an analytical instrument called Popsters, we have randomly selected and downloaded 90 user posts with complaints (30 posts per account), as of November 2020 and February 2021. We have chosen these months to avoid data distortion caused by public holidays. We have collected the comments for qualitative assessment, which included monitoring the responses to complaints in the comments to them, and also the institutional response to complaints by media and authorities beyond the social networking platforms.

Third, we also conducted in-depth interviews with three experts per region (30 experts altogether). For the interviews, the response rate was unexpectedly low (~20%). The questions posed discussed the types of accounts people prefer for lodging complaints, topicality of complaints, efficiency of government response, and roles of media in public response to citizens' complaints. The interviews, i.a., served for our decision to widen the list of regions and accounts for the main part of our study; as a result, we have enlarged the dataset more than two-fold. The interviews have also provided reasons to exclude Instagram from our final dataset for social network analysis of the discussions on complaints, as VK.com has shown to be significantly more salient contextually.

Fourth, at the final stage, when the main sample was formed, we have picked 63 accounts of media-like pages (accounts of traditional media and VK-based newsgroups) in 21 regions within the same periods. We have applied web crawling for gathering comments and metadata; the crawler had been previously developed by our working group [14, 33].

However, after the crawling stage less than 63 datasets were created. This was due to two factors: First, the crawler failed at several accounts; and second, some accounts returned a very small number of comments. We have additionally filtered out the accounts with the number of comments smaller than 25; this has left us with 42 accounts where the number of comments ranged from 40 to 1,726. For these accounts, we have reconstructed the web graphs with the algorithms of the Gephi library and assessed the graph centralities.

The accounts that were selected for the final dataset were of two types. The first group consisted of accounts of media of more traditional stance, like local newspapers, and the second type comprised newsgroups or accounts similar to 'overheard_[city]' (that grew out of humorous communities on what was overheard on public transport or local restaurants) and 'typical [city]' which described local events and small anecdotes.

3.2 Data Analysis

At the first stage of our research, we have qualitatively assessed the presence of complaints in posts and comments, monitored official reactions, and juxtaposed the experts' evaluations with the results of the analysis of complaints' content.

At the main stage, we have reconstructed the web graphs for each account and assessed the graph centralities, in particular the betweenness centrality and the PageRank centrality. Then, we have tried to find the media and political accounts among the commenting users and assess whether they played any important role within discussions. Then, we have applied correlational analysis, to see whether the discussion structure depends upon the number of users in it and the type of media (conventional/newsgroups). We see the discussion structure as either deliberative (a highly connected graph of an intense enough discussion) or cumulative, where the comments are individual and addressed to authors of the posts, not to fellow commenters; they accumulate in time but do not form a deliberative pattern of discussing.

To find connections between the type of media accounts (media/non-media), the structure of the graphs (deliberative/cumulative), the number of users and comments as potential third factors that shape the discussion structure, and the role of media and political accounts in the discussions, Spearman's rho correlations and the Mann-Whitney U tests were applied.

Due to the structure of data and the fact that access to Instagram was blocked in Russia in February 2022, our analysis was partly reshaped. This is why the stages of research do not always correspond to the logic of RQs and hypotheses. In the Results section, we will describe the outcomes of our analysis by hypotheses, not by the steps we have taken, first using the main dataset and then the preliminary dataset and expert interviews.

4 Results

RQ1. How Is the Discourse of Complaints Shaped on Social Networks?

H1.1. Comment conglomerates do not form deliberative communities in most cases. The dominant pattern of commenting is cumulative (represented by conglomerates of non-interlinked comments), not deliberative.

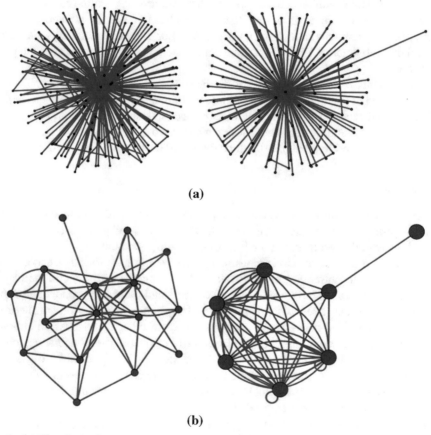

Fig. 1. (a) The discussion structure of cumulative (ego-network) type: Examples of web graphs for public pages 'ACT-54', Novosibirsk (vk.com/act54), and 'Yantarny DLB. Kaliningrad' (vk.com/amberbolt), left to right. (b) The discussion structure of deliberative type: Examples of web graphs for the public page 'Tupodar', Krasnodar (vk.com/typodar) and the media 'Belomorkanal', Arkhangelsk/Severodvinsk (vk.com/belomorchannel_tv29), left to right

This hypothesis was tested on the main dataset. We have, indeed, discovered two clearly different patterns of the discussion around complaints, which are represented via the web graph reconstruction in Fig. 1 (a, b).

As clearly seen from Fig. 1 (a, b), the first type of discourse is cumulative. The graph represents a final picture for each discussion where the users (nodes) comment on another account that has published the complaint. If stretched in time, the discourse looks like individual non-linked comments left by users and not discussed by other users. The resulting opinion is formed by accumulation of similar responses. The second type of discourse, on the contrary, is created by smaller numbers of users but is highly interconnected by reciprocal commenting. We observed 29 cumulative discussions and 14 deliberative.

To support our visual assessment of graphs with quantitative assessment, we have introduced two simple metrics for graph assessment:

- *Betweenness centrality ratio.* Betweenness centrality tells whether a particular node links other nodes in the graph. We have calculated mean betweenness centralities for the graphs, the main node excluded (the media that posted the complaint), thus assessing only the commenter nodes. Then, we have divided it by the number of users, to calculate the average betweenness centrality per node for each graph. The higher the ratio, the closer the pattern of a given graph is to the deliberative one.
- *PageRank centrality ratio.* PageRank centrality tells whether a given node is commented by other authoritative nodes in a discussion. We have calculated mean PageRank centrality for the graphs. Then, we divided the mean commenters' centrality by the main node's centrality, to see how big the difference is between the main node and the rest of users in the discussion. The lower the ratio, the closer the pattern of a given graph is to the deliberative one.

These mean centralities describe the graph in terms of deliberative/cumulative patterns. On Fig. 2 (a, b) we show that visual assessment of the graphs corresponds with the ratios, with only one exception in each case.

The Spearman correlations between visual assessment (coded as 1/2 as 'deliberative'/'cumulative') and the ratios were also strong (0.774** and 0.778** for the betweenness and PageRank centralities, respectively). This is why we see these ratio metrics as well-describing the deliberative vs. cumulative graph structure.

According to both the visual assessment and the combination of ratios, in our data, only 12 discussions of 42 demonstrated the deliberative pattern of discussing. This confirms H1.1 but not completely, as circa 28.5% of discussions still were of the deliberative type. This shows that VK.com as a platform can be a place for involving discussions around complaints, but not yet is for the majority of local media/newsgroups.

H1.2. *The deliberative pattern of discussing is lost with the growth of the number of commenters and comments in the discussions.*

To test H1.2, we have used the ratios calculated for H1.1. Then we have calculated Spearman correlations between the visual assessment ('type of graph') variable and the number of users, as well as Pearson correlations between the two ratio metrics, on one hand, and the number of users, on the other.

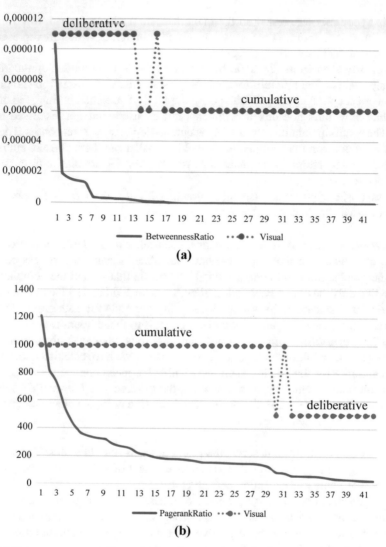

Fig. 2. (a) Betweenness centrality ratio vs. visual assessment of the accounts' web graphs. (b) PageRank centrality ratio vs. visual assessment of the accounts' web graphs

All the three variables have shown strong interdependence between the number of users in a discussion and the graph structure. The PageRank ratio has understandably shown higher correlation with the number of users than the visual assessment and betweenness ones (0.977** vs. 0.774** and 0.795**, respectively). However, as all the metrics have shown strong correlations, we can conclude that the number of users casts an impact upon the discussion pattern, and, with the growth of the number of participants, the deliberative pattern is lost. In our sample, the number of users in the deliberative discussions ranged from 40 to 125, and that in the cumulative discussions, from 173 to 1726, with just one exception of 115 users.

The same goes for the number of comments. The visually assigned type of graph and the two ratios show strong correlations with the number of comments ($0.778**$, $0.795**$, and $0.977**$, respectively). However, as the number of users strongly correlates with the number of comments ($0.965**$), assessing the number of comments does not add a lot to what we already know and does not allow for separate assessment of the role of comments with regard to the graph type. Thus, we have calculated the comment-per-user ratio for each graph. Here, correlations are much weaker: The visual graph type has no correlation to the comments-per-user ratio, while the two graph-based ratios weakly correlate with the latter ($0.314**$ and $0.452**$, respectively). This means that, even if users get engaged in commenting and leave multiple comments, they do not do it to the extent that it may influence the nature of discussions to make them more deliberative and interconnected.

This may be true for VK.com due to the fact that, in our sample, the number of comments per user is, in general, very low, which clearly tells of the cumulative character of commenting when users just 'drop' a comment or two and leave. As comment ratio does not correlate with the graph type, it means that this pattern is true for both 'deliberative' and 'cumulative' discussions. The average number of comments per user varies from 1.006 to 1.417, with just one exception of 2.137.

Thus, H1.2 is confirmed for the number of commenters and their respective comments, in case when the number of comments per user is low. When cumulative patterns of discussing are predominant, they get worse with the growth of the number of commenters and comments. On one hand, this goes against the pluralist view on the necessity of involvement of more people into discussing social issues and poses a range of questions on whether we as society need the discussions to be large, rather than small and focused. On the other hand, we clearly see that, even when the graphs demonstrate the deliberative character of the discussion, it must mostly depend on a small number of users, while the majority would only leave sporadic single comments.

H1.3. The type of media affects the discussion structure; media accounts differ from newsgroups in how the discussion is shaped.

We have divided the newsgroups where the complaints were posted into two groups: 1) the accounts of more traditional (often registered) media, like local newspapers, radio and TV channels, or web 1.0 news portals, and 2) the VK.com-based newsgroups, often of humorous stance.

Unlike the number of participants, the type of media in our data does not affect the graph type. The Spearman correlations between the 'type of media' variable and the variables that describe the graph structure (the 'visual assessment' one and the two ratios developed in H1.1) are all insignificant. We have also tested the potential difference between media and non-media accounts via Matt-Whitney U test, and it did not return any confirmed differences that would depend on media type – neither in the number of users nor any component of graph structure, including general betweenness and PageRank centralities, the centralities with the main node excluded, the centralities of main nodes, nor the two ratios suggested above.

This rejects H1.3, which, in its turn, supports the 'media decentering' thesis, as, in deliberative terms, conventional media do not produce or organize discussions of more deliberative nature than the web 2.0 newsgroups which are often far from any proper understanding of news production.

RQ2. What Is the Role of Media and Political Accounts in the Discussions around Complaints?

To address this research question with its three hypotheses, we have manually assessed the usernames within the collected data in each graph. First, we paid attention to the 10% of users with the highest PageRank and betweenness positions. Second, we had to assess the user lists on the whole, randomly checking manually the accounts which, as we would suspect, could belong to media, individual journalists, authorities, or their representatives. Thus, all the accounts were checked for usernames, and no less than 20% of them were also cross-checked on VK.com.

However, we have not found any single account that would recognizably belong to media, journalists, authorities as institutions, or civil servants. What we have discovered was a complete institutional and representative vacuum, even bigger than we had discovered for Twitter in 2017 [15].

Thus, we H2.1 and H2.2 cannot be answered directly. The answer for these hypotheses on dependence of media and authorities' roles on the deliberative/cumulative commenting or media type is that, for any type of discussion, institutions do not shape them to any extent, as they are absent from the discussions, and even the media accounts that post complaints do not participate in discussing them in their own accounts.

For H2.3, we have asked the experts on the roles of media in organizing the complaint – response mechanism. The experts have partly confirmed that media are decentered and are more and more excluded from the complaint resolution mechanism. Instead, the experts underlined the growing role of systematic complaint management activities recently established within the Regional Management Centers (TSURs) of regional governments. These activities have been automatized via a system of automated complaint collection from social networks and web 1.0 portals; this system is called 'Incident Management' and is seen as key to resolving complaints. However, 50% percent of the experts stated that independent, socially-mediated news accounts were the main space where users complained, while authorities' accounts were named by 28.9%, and conventional media by only 15.7%.

Thus, the role of media as organizers of discussions around complaints is limited to their 'spatial' dimension, to their role of a forum for complaints. H2.3 is confirmed for newsgroups in terms of their role as spaces for venting users' emotions via complaining, but not for media or media-like accounts as actors that help shape meaningful discussion upon these complaints. Our conclusion may contribute to the concept of journalism decentering in terms of media roles on social networks: One of the media functions, namely the 'spatial' one, still works, but other ones, including organization of public discussion, are not performed within the networked discussions.

RQ3. How is the Institutional Response to Complaints Organized Instead?

To answer this question, we have used the data from the first stage of our research, where we employed a smaller number of accounts but also looked at Instagram, as well as asked experts on the complaint – response chains.

What we monitored on this stage was not only the institutional response on social media but also their external response. By the latter, we mean publication of responses to complaints on governmental websites and media publications (mostly news).

We have discovered that the complaints *are* responded to in the public sphere beyond networked discussions (see Fig. 3). The discovered percentage of complaints addressed varies highly by region, and there are no macro-regional patterns for how intensely the public sphere actors respond. Thus, regional variances demand further research on factors of (non-)answering.

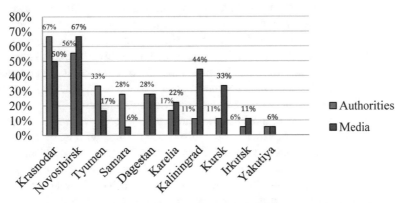

Fig. 3. The percentage of responses by the authorities and media, by region

Our main conclusion for RQ3 is that the institutional actors of the regional public spheres use their conventional instruments of response to complaints. Instead of engaging into open discussions on social networks, the authorities report on the resolved complaints on their websites, while media pick up complaints as a source for their agendas and respond via publishing news, thus returning to the practice of collecting public problems via social media. However, in the socially-mediated public spheres, it turns the chain of 'public dissent – media alert – governmental response' into a fork-like scheme where media and authorities pick up the complaints simultaneously (or, in some cases, authorities may even outperform media in terms of speed and completeness of response). This, again, adds much to decentering media in the public space, while also diminishing the opportunities for local communities to participate in substantial discussions on the issues raised. The complaints communication appears to be one-way again, even if this way is not 'authorities to people' but 'people to authorities.' The feedback loop from the public sphere in response to complaints is there, but is not involving.

5 Discussion and Conclusion

In Russia, as in the rest of the world, the role and tasks of traditional media are changing. Not only they are increasingly fading into the background and pushed aside by new media platforms such as social networks, but they are also being transformed by the influence of social networks. Local news accounts cease to be information suppliers only; given the peculiarities of the Russian political and information environment, nor can they become a representation of the political will of the regional community. Ac-quiring the technical qualities of new media, they instead become a place of concentration of the negative emotions of citizens, of their complaints and discontent. In some sense, city news accounts on social media have become a 'place' where residents of small towns feel 'at home' sharing their emotions and cheering each other up – that is, a 'domesticated' part of the public sphere. On one hand, however seemingly small, this function of hyperlocal news accounts cannot be overestimated, as they have created an arena for bottom-up policy criticism that is truly rarely found in post-Soviet regimes.

We have discovered that the practice of user complaints via local news accounts on social networks has become widespread in Russia, and the local authorities and media do pay attention to relatively small-size complaints by individual Internet users. This practice is still new, and the reaction by neither media nor civil servants shows stable patterns in terms of volume of attention and efficiency of complaints. In any case, for Russia, we see how Internet platforms have become home to an emergent practice of direct, de-mediatized, and network-facilitated public accountability.

On the other hand, local authorities, more than professional journalists, help ensure that discontent begins and ends in the space of social networks, without going beyond it. A response to a complaint beyond the socially-networked discussion itself may become a substitute for solving the problem in reality, or at least serves as a way to end the discussion on a given complaint by answering outside its place of initial appearance.

We have shown how the practices of institutional reaction to complaints decenter local media, putting them into competition for decision-making agendas with the author-ities who use the 'Incident Management' monitoring system and are sometimes quicker to respond. The absence of media in the VK.com discussions on complaints leaves them with the only function of space provision for venting discontent and dissent, but pre-vents media from performing the functions of discussion organizers and watchdogs. New types of media-like accounts successfully compete with traditional editorial offices in popularity as complaint spaces, but neither they help involve institutional actors into two-way discussions on local problems.

Acknowledgements. This research has been supported in full by the project 'Center for International Media Research' by St. Petersburg State University, project #92564627.

References

1. Almond, G.A., Powell, G.B.: Comparative politics: A developmental approach. Little Brown, Boston, MA (1966)
2. Jangdal, L.: Local democracy and the media: can hyperlocals fill the gap? Nordicom Review **40**(s2), 69–83 (2019). https://doi.org/10.2478/nor-2019-0027
3. Dovbysh, O.: New gatekeepers in town: how groups in social networking sites influence information flows in Russia's provinces. Soc. Media Soc. **7**(2), 1–11 (2021). https://doi.org/10.1177/20563051211013253
4. Zubarevich, N.: Chetyre Rossii (Four Russias). Vedomosti, 30 Dec 2011 (2011). https://www.vedomosti.ru/opinion/articles/2011/12/30/chetyre_rossii

5. Bodrunova, S.S.: Practices of cumulative deliberation: a meta-review of the recent research findings. In: Chugunov, A.V., Janssen, M., Khodachek, I., Misnikov, Y., Trutnev, D. (eds.) EGOSE 2021. CCIS, vol. 1529, pp. 89–104. Springer, Cham (2022). https://doi.org/10.1007/978-3-031-04238-6_8

6. Schroth, F., Glatte, H., Kaiser, S., Heidingsfelder, M.: Participatory agenda setting as a process — of people, ambassadors and translation: a case study of participatory agenda setting in rural areas. Euro. J. Futures Res. **8**(1), 1–12 (2020). https://doi.org/10.1186/s40309-020-00165-w

7. Coleman, S., Freelon, D.: Handbook of digital politics. Edward Elgar Publishing (2015). https://doi.org/10.4337/9781782548768

8. Bodrunova, S.S.: Contributive action: socially mediated activities of Russians during the COVID-19 lockdown. Media Int. Aust. **177**(1), 139–143 (2020). https://doi.org/10.1177/1329878X20953536

9. Nielsen, R.K.: News media, search engines and social networking sites as varieties of online gatekeepers. In: Rethinking Journalism Again (pp. 93–108). Routledge (2016)

10. Gunarathne, P., Rui, H., Seidmann, A.: Whose and what social media complaints have happier resolutions? Evidence from Twitter. J. Manage. Inf. Syst. **34**(2), 314–340 (2017). https://doi.org/10.1080/07421222.2017.1334465

11. Hogreve, J., Eller, T., Firmhofer, N.: When the whole world is listening – an exploratory investigation of individual complaints on social media platforms. In: Bruhn, M.; Hadwich, K.; Eds. Dienstleistungsmanagement und Social Media, Springer Gabler, Wiesbaden, Germany, pp. 515–540 (2013). https://doi.org/10.1007/978-3-658-01248-9_23

12. Swart, J., Groot Kormelink, T., Costera Meijer, I., Broersma, M.: Advancing a radical audience turn in journalism fundamental dilemmas for journalism studies. Digital Journalism **10**(1), 8–22 (2022)

13. Wahl-Jorgensen, K.: News production, ethnography, and power: on the challenges of newsroom-centricity. In: Bird, S.E. (ed.) The Anthropology of News and Journalism: Global Perspectives. Indiana University Press, Bloomington (2009)

14. Bodrunova, S.S., Litvinenko, A.A., Blekanov, I.S.: Please follow us: media roles in Twitter discussions in the United States, Germany, France, and Russia. Journal. Pract. **12**(2), 177–203 (2018)

15. Smoliarova, A.S., Bodrunova, S.S., Blekanov, I.S.: Politicians driving online discussions: are institutionalized influencers top twitter users? In: Kompatsiaris, I., et al. (eds.) INSCI 2017. LNCS, vol. 10673, pp. 132–147. Springer, Cham (2017). https://doi.org/10.1007/978-3-319-70284-1_11

16. Bodrunova, S., Smoliarova, A., Achkasova, V., Blekanov, I.: Who is to blame? Patterns of blaming and responsibility assignment in networked discussions on immigrants in Russia and Germany. J. Soc. Policy Stud. **16**(4), 627–644 (2018)

17. Chen, J.: Useful complaints: how petitions assist decentralized authoritarianism in China. Lexington Books: Lanham, MD, p. 204 (2016)

18. Stockmann, D., Gallagher, M.E.: Remote control: how the media sustain authoritarian rule in China. Comp. Pol. Stud. **44**(4), 436–467 (2011)

19. Toepfl, F., Litvinenko, A.: Transferring control from the backend to the frontend: a comparison of the discourse architectures of comment sections on news websites across the post-Soviet world. New Media Soc., pp. 1–18 (2017). https://doi.org/10.1177/1461444817733710

20. Bodrunova, S.S., Blekanov, I.S., Maksimov, A.: Public opinion dynamics in online discussions: cumulative commenting and micro-level spirals of silence. In: Meiselwitz, G. (ed.) HCII 2021. LNCS, vol. 12774, pp. 205–220. Springer, Cham (2021). https://doi.org/10.1007/978-3-030-77626-8_14

21. Bodrunova, S.S., Blekanov, I.S.: A self-critical public: cumulation of opinion on Belarusian oppositional YouTube before the 2020 protests. Soc. Media Soc. **7**(4), 205630512110634 (2021). https://doi.org/10.1177/20563051211063464

22. Toepfl, F.: Comparing authoritarian publics: the benefits and risks of three types of publics for autocrats. Commun. Theory **30**(2), 105–125 (2020)
23. Toepfl, F., Litvinenko, A.: Critically commenting publics as authoritarian input institutions: how citizens comment beneath their news in Azerbaijan, Russia, and Turkmenistan. J. Stud. **22**(4), 475–495 (2021). https://doi.org/10.1080/1461670X.2021.1882877
24. Litvinenko, A., Toepfl, F.: The (non-)adoption of participatory newsroom innovations under authoritarian rule: how comment sections diffused in Belarus and Azerbaijan (1998–2017). Digit. Journal. **6**(7), 1–22 (2021). https://doi.org/10.1080/21670811.2021.1888137
25. Internet control through ownership: the case of Russia. Post-Soviet Affairs **33**(1), 16–33 (2017). https://doi.org/10.1080/1060586X.2015.1121712
26. Bodrunova, S.S.; Litvinenko, A.A.: Four Russias in communication: fragmentation of the Russian public sphere in the 2010 s. In: Dobek-Ostrowska, B., Glowacki, M.; Eds. Democracy and media in Central and Eastern Europe 25 years on, Peter Lang, Frankfurt am Main, pp. 63–79 (2015). https://doi.org/10.3726/978-3-653-04452-2
27. Bodrunova, S.S., Litvinenko, A., Blekanov, I., Nepiyushchikh, D.: Constructive aggression? Multiple roles of aggressive content in political discourse on Russian YouTube. Media Commun., 9, 181–194 (2021). https://doi.org/10.17645/mac.v9i1.3469
28. Litvinenko, A., Toepfl, F.: The "gardening" of an authoritarian public at large: how Russia's ruling elites transformed the country's media landscape after the 2011/12 protests "For Fair Elections." Publizistik **64**(2), 225–240 (2019). https://doi.org/10.1007/s11616-019-00486-2
29. Litvinenko, A., Nigmatullina, K.: Local dimensions of media freedom: a comparative analysis of news media landscapes in 33 Russian regions. Demokratizatsiya: The Journal of Post-Soviet Democratization **28**(3), 393–418 (2020)
30. Nigmatullina, K., Rodossky, N.: Pandemic discussions in VKontakte: hopes and fears. In: Meiselwitz, G. (ed.) HCII 2021. LNCS, vol. 12775, pp. 407–423. Springer, Cham (2021). https://doi.org/10.1007/978-3-030-77685-5_30
31. Mediascope WEB-Index. https://webindex.mediascope.net/report/general-statitics?byGeo= 1&byDevice=3&byDevice=1&byDevice=2&byMonth=202005&id=16571&id=88155& id=156688&id=12808. Accessed 26 July 2021
32. Smoliarova, A., Bodrunova, S.S.: InstaMigrants: global ties and mundane publics of Russian-speaking bloggers with migration background. Soc. Media Soc. **7**(3), 205630512110338 (2021). https://doi.org/10.1177/20563051211033809
33. Nepiyushchikh, D., Blekanov, I.: Data crawling approaches for user discussion analysis on web 2.0 platforms. In: Smirnov, N., Golovkina, A. (eds.) SCP 2020. LNCISP, pp. 793–800. Springer, Cham (2022). https://doi.org/10.1007/978-3-030-87966-2_91

Comparative Analysis of Topics Covered by False and True News in the Context of the COVID-19 Pandemic

Margarita Zhdankina[✉] [ID], Victoria Kolesnikova[✉] [ID], Sergey Romanov[ID], and Dmitri Rudyuk[ID]

HSE University, Moscow, Russia
mzhdankina@hse.ru, vekolesnikova@edu.hse.ru

Abstract. The COVID-19 pandemic has been an urgent topic of discussion in various media and social networks for a couple of years already. The lack of research and the rapid spread of the virus around the world only contribute to an increase in media interest in this topic, but these same reasons contribute to the emergence of many fake-news headlines and conspiracy theories. The article presents a study of the media coverage of the COVID-19 pandemic and the language used in the framework of fake news and its comparison with real news. The concept of 'fake news' does not have a single definition, but within the framework of this study, intentionally unreliable media reports, including propaganda, would be considered as such. Not only the spread of unreliable information by fake news but the undermined trust in news and media institutions are highlighted in the literature as the main alarming factors. With the help of network analysis, the connections between the words most intensively used in fake and real news headlines were identified. An analysis of correlations between the most frequently mentioned words in the headlines was carried out, based on which networks were built to analyze the main topics of news related to COVID-19. The study was conducted in an exploratory format and can be used as a basis for a deeper analysis of fake news related to the COVID-19 pandemic. In order to compare what drives words to form new phrases in both type of news headlines, the ERGM models were implemented with usage of BING sentiment variable and two structural network variables.

Keywords: COVID-19 · Media · Fake news · Network analysis · Network agenda

1 Introduction

COVID-19 has been an integral part of almost every person's life for several years now. The new infection, which has captured all continents so rapidly, of course, could not help but cause a vivid media response to its appearance. Therefore, news and rumors about this virus began to spread through all possible information channels long before the first cases of infection became known outside of China. The novelty of this disease

A. Antonyuk and N. Basov (Eds.): NetGloW 2022, LNNS 663, pp. 21–35, 2023.
https://doi.org/10.1007/978-3-031-29408-2_2

has contributed to the emergence of numerous interpretations of what is happening, as well as conspiracy theories. It should be noted that a significant role in constructing an understanding of the pandemic phenomenon is played by the media. They present a discussion of ways to protect oneself from the disease or recover from it, as well as many other topics related to COVID-19.

The problem of covering the pandemic of the new coronavirus infection is extremely acute. In the face of a lack of reliable information, the phenomenon of 'fake news' spreading both through media outlets and social networks has begun to manifest itself on a large scale. The abundance of false information leads to increasing anxiety in society, as well as misunderstanding of the nature of the pandemic and how one should behave to be safe in the current conditions [15].

In this paper we will analyze the topics covered in fake news headlines based on word usage in them and compare them with the titles of reliable articles. In our opinion, distinguishing the difference between two types of news can help with the further development of approaches to identifying fake news, including the improvement of machine learning models, and, potentially, the development of ways to counter them in media.

2   Theoretical Framework

2.1   'Fake News' Phenomenon

The problem of fake news is not new, in particular, it was raised in the context of the 2016 US presidential elections, where both sides widely used to spoil the rival's reputation built, among other things, on unreliable news [1]. Of course, the events of the COVID-19 pandemic have also become fertile ground for the spread of fake information. In turn, such heated topic has led to the spread of biased narratives fueled by unsubstantiated rumors, misleading news, and paranoia [9]. But, according to Bryanov and Vzyatishcheva, the main danger of fake news lies not in deceiving the audience, but in undermining confidence in news media as a social institution [4].

First of all, it should be noted that researchers most often understand 'fake news' as various forms of intentional misinformation, such as false facts, data manipulation, and propaganda. At the same time, rumors, misguided interpretations of facts, and other unintentionally unreliable media reports are excluded from this concept [11]. Although fake news can be spread both through publications and through social networks, the focus of our study is on fake news published in media intentionally.

The phenomenon of fake news often goes hand in hand with propaganda. The concept of propaganda is often associated with politics, however, in general, the mechanism of influence (regardless of the topic) is based on the distortion of information or its suppression to promote a certain ideology [27]. That is, propaganda specifically produces fake news to achieve a certain result, for example, in elections, as well as to manipulate public opinion [27]. However, propaganda is far from the only possible reason for the appearance of fake news, since conspiracy theories also contribute to the formation of fake news. Some of them, which arose against the backdrop of the COVID-19 pandemic, continue to exist to this day and gain a huge number of followers [12]. Therefore, in this study, we will not focus specifically on the concept of propaganda in the context of

COVID-19 but will consider news related to coverage of the pandemic, regardless of the context or method of influencing the audience.

Moreover, the appearance of fake news may depend on the concept of relexicalization. Based on Trew's research 'Linguistic variation and ideological difference' [14], we consider that relexicalization is a process of changing the linguistic features for constructing phrases (in other words, replacing some words with synonyms and changing word order), especially news headlines, which may cause changes in perception of news content. In the context of fake news, relexicalization can lead to the fact that even a reliable headline, due to the author's interpretation of the presentation, can lead to a distortion of the audience's perception of the event and, thus, also fall under our definition of fake news.

As Dant outlined, relexicalization might be observed in the discourse of several idcologically opposed groups [10]. In our research we do not focus on the groups of interest, but mainly on language, using an annotated and preproccsscd data set. Our particular focus is to examine language features of fake and reliable headlines, to analyze how words within them connect to each other, and to inspect the difference.

2.2 Network Agenda Setting Theory

Back in 1922, Lippmann suggested that media be perceived as something that 'builds a bridge between the real world, 'the world outside', and the picture in our head' [22]. The study of fake news seems to be relevant precisely in connection with the influence that the media can have on public opinion and people's perception of the significance of certain problcms [23]. However, for our study, the theory of the agenda of the third level [17] is more significant, which moves away from the binary opposition of media/agenda to a more relational approach. This may explain the structural essence of our main research objects – words and phrases in news titles. We are looking at the syntagmatic of headlines because words get their characteristics based not on their lexical features but from their place in a network of the words. In our case this means that building a network not on phrases, but on words, as structural approach is a relevant method of analyzing the studied objects. The word is a unit of language, its elementary component. In other words, it is indivisible into smaller segments as regards its nominative function. Words in general resist interruption. That means, we cannot freely insert pieces into words as we do into phrases or sentences. Therefore, for the syntagmatic of headlines, we require the analysis of their parts – that is, to the network analysis of the words they consist of.

Guo and McCombs proposed to consider the media and public opinion not just as two spheres influencing each other, but as interrelated networks of news items [16]. In other words, the mutual relations of news items in the structure of the media in one way or another determine the structure of the public agenda. This approach focuses more on the relationships between nodes (news items) in networks than on the nodes themselves (see Fig. 1).

Each news item, which can be a separate politician [18] or some kind of event (war, economic crisis, etc.) [15], occupies the position of a node in the network. In the mentioned articles that used this theory, the most meaningful results were obtained by comparing two or more networks. Thus, network agenda setting theory allows us to use the network methodology for analyzing media headlines and look for the networks

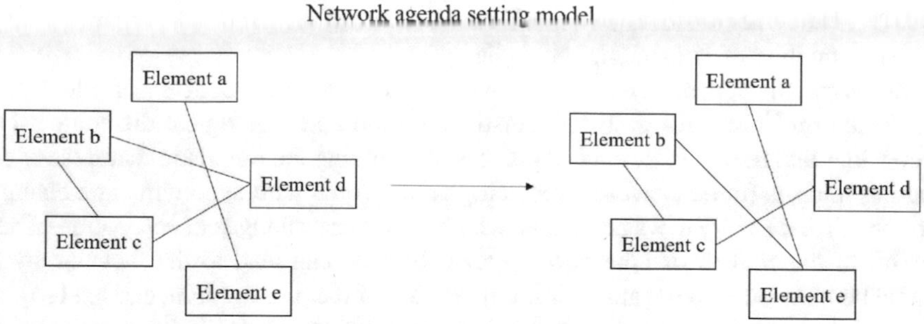

Network agenda setting model

Fig. 1. Network agenda-setting model. Figure adapted from [17]

of words in them. Moreover, third-level agenda approach allows implementing diverse comparative studies. In this paper, we will consider networks of fake and reliable news headlines.

In our work, we relied on a study based on publications containing fake news about a new coronavirus infection in six countries in Latin America [5]. This paper has helped us to understand how to pre-process data for the network analysis of covid-related publications in media. In addition, the idea of using words as a unit of analysis was also derived from this research. During their analysis, the authors identified significant similarities between the keywords in publications from different countries using network analysis. Thus, we pose the main research question of this work as: do fake and reliable news and their phrase formation differ, and how?

3   Methodology

3.1   Database Description

The main criterion for choosing a database for analysis was whether the data belonged to the time period of interest, therefore we chose the 'COVID-19 Fake News Dataset' database [8]. This database is aggregated using Webhose.io data relating to the period December 2019–July 2020.

We chose this period to study the key topics of news about COVID-19 at the beginning of the pandemic when the first waves of the spread of the disease were noted. This dataset was chosen for analysis to avoid considering the stereotypical themes that appeared in the media in the process of adapting to the virus topic and to see how society and the media reacted to the new emerging danger. Since in the case of discussing a long-term virus, the media can move from constructing new meanings to spreading previously existing ones, which in the framework of our study may lead to a shift of the data obtained into one of the topics.

The data includes the titles and full texts of news articles, which is pre-marked as 'true' and 'false', and the category 'partially false' within this encoding is included in 'fake'.

In total, there is 3119 news in the database, of which 2061 are true, and 1058 are false or partially false. In our analysis we focus only on article titles, not the whole text of them.

The disadvantage of this database is the lack of indication of the source of particular news. However, in the case of analyzing fake news, the specific media from which the headline or text was taken is not so important.

3.2 Data Processing and Analysis Steps

The most frequently mentioned words in the headlines of fake news about COVID-19 were various names of the virus itself: 'corona', 'virus', 'covid', 'ncov', 'coronavirus'. Those words were deleted from the database, because potentially they could be the keywords for the data search and extraction. For preprocessing and analysis of data, we used R [24].

The first step in data preparation was to remove words mentioned less than 10 times in each set of news titles to find topics that are relevant in the context of the coronavirus discussion in the text media. Based on the obtained results, the subsequent operations of preparing and building words networks were carried out. To map ties between words, we calculated correlations of pairs of words in titles of fake and reliable news using 'widyr' package [25]. We used phi coefficient which represents a measure for binary correlation. Similar to Pearson correlation coefficient, phi coefficient takes on values between −1 and 1. It shows the likelihood that some words co-occur in the same title rather than in different titles (the closer the coefficient is to 1, the higher is the correlation or the chance of word co-occurrence). Next, for both types of titles, we selected 100 word pairs with highest correlations. We had to limit the number of correlations used for visualization, since adding more connections could lead to unreadability of the resulting networks. Using the obtained word pairs and correlation coefficients, 'fake' and 'real' networks were constructed with 'ggraph' package [28]. Finally, we visualized the resulting networks. The visualizations reflect main patterns of paired formations of words and tie strength. Thus, information was obtained about the emerging mutual mentions of words and about how likely they are to be used together in news headlines.

Then, we added the node sentiment variable to the networks which was constructed with BING sentiment lexicon [3]. We implemented lexicon-based sentiment analysis of our corpora: words from both 'fake' and 'true' data were matched with words from BING − vocabulary of general-purpose English sentiment lexicon which categorizes words in a binary manner, either positive or negative. In comparison, AFiNN lexicon has continuous sentiment metric, but in our work, we only use it to demonstrate the limitations of sentiment analysis [13]. Words which have not found a match were categorized as 'neutral' and were put in the control group of nodefactor predictors. Such strategy for sentiment analysis was chosen due to its simplicity and the relevant nature of the BING lexicon, as it was developed based on the texts from social media, news, reviews, forum discussions, and blogs [3]. Another structural variable is degree centrality which is defined as the number of ties incident upon a node and can be interpreted as the 'popularity' of a node in a network.

Before resuming, it is important to note some limitations of such sentiment analysis. As sentiment of the lexicon comes from pretrained models there might be errors with connotations of different words due to the lack of contextual sentiment in those lexicons. For example, lexemes 'positive covid' and 'positive test' were marked with a good and positive connotation in the AFINN lexicon which gives an ambivalent result and restricts the modeling (see Table 1), so we have not used continuous encoding for our features.

Table 1. Words from Covid dataset with sentiment points from AFINN lexicon

Word 1	Word 2	Correlation	AFiNN value
death	tall	0.474	-2
infected	candidate	0.337	-2
infected	antibodies	0.308	-2
protect	reports	0.279	1
infected	patients	0.224	-2
death	dr	0.209	-2
flu	death	0.194	-2
death	flu	0.194	-2
flu	dr	0.189	-2
positive	*covid*	*0.188*	*2*
positive	test	0.182	2
infected	vaccine	0.180	-2
emergency	disease	0.165	-2
positive	patient	0.151	2
emergency	treatment	0.151	-2
infected	ncov	0.140	-2
emergency	test	0.135	-2

As a result, we built networks for both fake and real news. The method of constructing all networks was identical for both types of information messages so that it was possible to subsequently compare the obtained results.

As for the second step, we used exponential random graph modelling (ERGM) to examine factors that might affect tie formation in the networks. Furthermore, we built models based on the top 100 word pairs, since there were very weak correlations in the fake news network, which would make the analysis unreliable. For the correct comparison of models, the top 100 word pairs were also used in the reliable news network. For both 'fake' and 'true' networks we reduced the tie strength values to the presence/absence of the tie. Then, we used several variables – BING node sentiment and several structural variables. We have built 3 models for 'fake' and 'reliable' news with the gradual addition of degree, GWESP (geometrically weighted edgewise shared partner) and sentiment variables to identify their impact on the formation of headlines. It should be noted that to build the models we used the database before deleting the search keywords. This is important so that models on different networks can be compared with each other, since keywords are necessary centres for building new connections and determining the probability of their construction.

4 Fake News Analysis

Analysis of changed data revealed a group of words associated with the place where the first outbreaks of coronavirus were reported – 'China', 'Wuhan' (see Fig. 2). Such results can indicate that the chosen data covers the period of the emergence of new infection. This refers to time period when the world media discussed it only as part of the outbreak in Wuhan. Despite the fact that the infection did not have time to spread to other countries, interest was relatively high already.

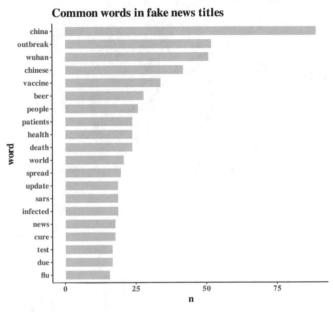

Fig. 2. The most frequently mentioned words in the titles of fake news articles

In the course of analysis of word relationships, namely their joint mentions in news titles, it was found that most headlines are concentrated on the topics covering morbidity and vaccination. This connection of the most frequent words is visible in the cluster consisting of the words 'infected', 'antibodies', 'patients,' 'vaccine' and 'candidates' (see Fig. 3). Another cluster could presumably represent the topic of the impact of the new disease on the market: a connection can be traced between the words 'impact', 'market', 'report' and 'sars'. An important tie located further from the biggest cluster of joint mentions is 'health'–'government', which may be a reflection of the headlines mentioning government decisions regarding the preservation of public health.

The strongest connections, in addition to the combination of 'corona' and 'virus', are observed between the words: 'vaccine' and 'candidate', 'antibodies' and 'candidate', 'infected' and 'candidate', 'market' and 'impact', 'death' and 'toll' (see Fig. 4). From the results, it becomes clear that news headlines are geared towards clickbait or sensationalism to attract more attention from readers.

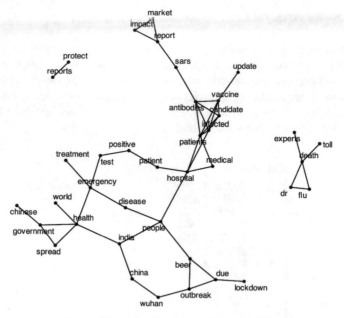

Fig. 3. Network of words in fake news titles

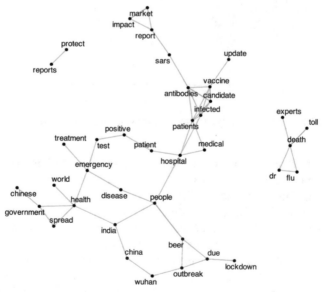

Fig. 4. Tie strengths between words in fake news titles. Tie color indicates tie weights: green represents tie with more weight.

5 Analysis of Reliable News

The most frequently mentioned words in the headlines of real news about COVID-19, as well as in fake messages, were the fact of the virus emergence: the country of origin, town 'Wuhan' and the globality of spreading. In general, the names of reliable news have approximately the same number of mentions of various words related to the pandemic. Occasionally, new concepts appear, such as 'emergency' or 'confirmed' (see Fig. 5).

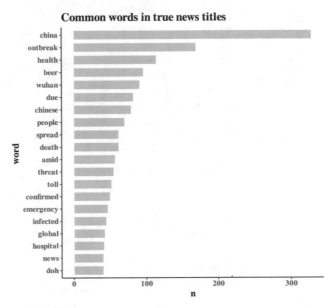

Fig. 5. The most frequently mentioned words in the titles of real news articles

However, despite the similar pattern of word mentions, their co-occurrence in headlines differs from the pattern we observed in the case of fake news. Ties 'epidemic'–'control' and 'travel'–'ban' (see Fig. 6) are emerging, which indicate news coverage not only of the spread of the disease, but also of methods used to slow down the transmission of the disease. The phrase 'preventive' – 'measures' can also indicate the focus of reliable news on measures of preventing the spread of the virus.

Unlike fake news headlines, reliable news titles have an average correlation between words related to the study of the disease and information about it. The global character of the event is also important when creating article titles, which is reflected in the relationship between 'global', 'emergency' and 'declares' (see Fig. 7).

An interesting point is that the network of words in the headlines of reliable news is weakly connected. There are many words related exclusively to each other, for example, 'negative' – 'tested', 'travel' – 'ban' (see Fig. 7). The appearance of more separate phrases may indicate the coverage of specific events, which is exactly what distinguishes real news. Covering more diverse topics leads to the emergence of such separate pairs of words, in contrast to fake news headlines. The latter use a single limited set of topics

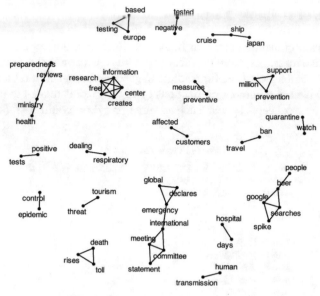

Fig. 6. Network of words in real news

Fig. 7. The strength of connections between words in real news. Tie color indicates tie weights: green represents tie with more weight.

and terms to attract the attention of readers, since the coverage of events that took place on the world stage is not their priority.

6   ERGM

To substantiate previous observations, we seek a textual model which could help us understand the difference in phrase formations of our networks. Earlier research on socio-semantic networks has come up with innovative methods for examining text data, amongst them there is a wide usage of various topic modeling methods, including in combination with network-based models [6, 7]. However, these studies are not focused on the mechanisms that operate between words themselves, which is what we consider necessary for our complete analysis.

Thus, we reduced these approaches to the level of words, and combined them with the ERGM approach, which has already demonstrated its performance on word-level socio-semantic networks [2]. We used the 'ergm' package for R [21], a suitable statistical tool implementing a particular class of models – ERGMs or p-star models [20]. Such models can provide statistical inferences regarding processes underlying network formation – in our case, words connecting into phrases. For the start we have built two unconditional models, or Erdös-Rényi models (see Table 2), using only the edges predictor, following the baseline procedure for ERGM implementation [21].

Table 2. The unconditional models

	Model 1 ('fake' news)		Model 2 ('reliable' news)	
	EST	SE	EST	SE
Structural features				
Edges	**–2.5696**	0.146	**–3.3945**	0.143
AIC	362.7		443.1	
BIC	367.2		448.4	

Note: Coefficients in bold are significant at the 95% level

Then, ERGMs were created with non-parametric degree predictors (standing for the number of ties) for both datasets (see Models 3 and 6 in Tables 3 and 4, respectively). These models account for potential influence of the overall popularity of a word on the probability of word pairing. The coefficients for degree (1) centrality were partly significant for the network of 'reliable' news, as well as for the 'fake' news network. However, it was illogical, from the first glance, because in the denser 'fake' news network one might expect a statistical significance from the degree predictors of a higher level.

These counter-intuitive outcomes made us search for a comprehensive explanation and another predictor. As a result, we obtained a significant result from GWESP structural predictor (see Models 4 and 7 in Tables 3 and 4) for the 'fake' and 'reliable' news networks. GWESP is a dyad-based configuration, rather than node-based, that accounts for triadic closure [19, 26]. However, this parameter disables the usage of degree (1), because technically the former is the linear combination of the latter statistic and additional information about nodes partnership.

Table 3 The models for the network of 'fake' news

	Model 3		Model 4		Model 5	
	EST	SE	EST	SE	EST	SE
Structural features						
Edges	**−1.832**	0.396	**−3.771**	0.297	**−3.749**	0.299
Degree (1)	**2.119**	0.941				
Degree (2)	1.118	0.780	0.518	0.433	0.540	0.418
Degree (3)	1.206	0.620	**1.046**	0.408	**1.079**	0.392
GWESP			**1.141**	0.201	**1.140**	0.195
Sentiment features						
Positive BING sentiment					0.044	0.231
Negative BING sentiment					− 0.211	0.383
AIC	360.2		331.7		334.7	
BIC	378.4		350		362	

Note: Coefficients in bold are significant at the 95% level

Table 4. The models for the network of 'reliable' news

	Model 6		Model 7		Model 8	
	EST	SE	EST	SE	EST	SE
Structural features						
Edges	**−2.481**	0.380	**−4.220**	0.236	**−4.300**	0.548
Degree (1)	**2.371**	0.768				
Degree (2)	1.248	0.648	0.260	0.322	0.250	0.309
Degree (3)	0.484	0.602	− 0.154	0.451	− 0.154	0.465
GWESP			**1.405**	0.225	**1.423**	0.221
Sentiment features						
Positive BING sentiment					0.035	0.277
Negative BING sentiment					0.107	0.433
AIC	427		406		409.9	
BIC	448.4		427.4		441.9	

Note: Coefficients in bold are significant at the 95% level

This statistic showed a significant result for both networks, but, according to Models 4 and 7, its combination with centralities leads to different significance of the degree predictors. Whereas in Model 7 for the 'reliable' network their significance disappears, in Model 4 for the 'fake' network it is significant at the level of degree (3), so the probability of tie establishment in the 'fake' network is statistically dependent not only on triadic closure, if it is introduced, but also on node popularity.

We interpret it in such way: for 'fake' and 'reliable' networks, the more closed triads are already formed, the more probable that triad would form another tie. If two words in the phrase are both used with a third one, they have higher chances of forming a new phrase. Nevertheless, for the former network the number of co-occurrences of a word with other words still affects tie formation if effect of GWESP is added to a model. One can refer to the picture of the 'fake' network, which demonstrates that it has a branch of words to which we refer in the Sect. 4 – a sort of a thesaurus of 'fake news' in our corpus. So, the presence of a word in that set increases its probability of becoming even more popular.

Furthermore, we created models with exogenous nodal attributes formed by referring words from our data set to the sentiment lexicon databases of BING [3]. The influence of sentiment attributes was assumed, for the more emotional impulse the word gives, the higher, supposedly, its probability to form phrases. Nonetheless, the statistical results showed that this supposed impact is insignificant (see Models 5 and 8 in Tables 3 and 4). We assume that our hypothesis was based more on sentiment, than on the structural part of language, because the sentiment databases use only a few non-specified and general connotations.

In a nutshell, the ERGMs showed that predictors of tie formation have opposite effect on each other in different networks: Whereas GWESP in the network of 'fake' news makes degree centralities significant, in the 'reliable' news degree centralities lose all their significance when it is used. GWESP does not allow building models with degree (1), therefore it is impossible to talk about changes in its significance due to its absence. However, GWESP does not add significance to degree (2) and degree (3), respectively, centrality loses its significance in the model.

Such results outline that in fake news headings a word connects to another if it is related to a particular set of nodes limiting the network and the word connection. Some words in unreliable titles constitute a thesaurus used for the Covid-19 topic – a branch which could be observed on the Fig. 4, – and if the word is not present in this thesaurus, it is less probable to form a new phrase.

For 'reliable' news, word popularity is not significant, which means that co-occurrence will not go beyond the triads. This is confirmed by the co-occurrence visualization in Fig. 7.

7 Conclusion

In this study we performed a comparative analysis of real and fake news using methods of network analysis and had some noticeable results. First of all, it turned out that the headlines of fake news have less thematic diversity compared to headlines of the real ones. Publications that specialize on fake news focus more on the emotional aspect of the titles than on informativeness and verbatility. Also, there is frequent use of the most relevant and prominent keywords that most readers, regardless of their level of knowledge on the topic, will be familiar with (perhaps this may also be due to the focus on SEO algorithms).

Other important results were obtained from ERGMs. According to ERGMs, 'fake' news network has a specific feature: the significance of degree centralities and GWESP suggests that the more closed triads are formed, the more probable is formation of another tie. We interpret this as there can be a sort of a thesaurus of 'fake news' headlines' keywords with which other words tend to connect more than with others. In comparison, in 'reliable news' degree centralities are not significant. It means that the number of word's ties does not affect the potential co-occurrence of words in the heading of true news, and we cannot claim that in reliable headlines words are affected by their 'popularity'.

Our study has a limitation related to the fact that people involved in determining the degree of reliability of the information, 'fact checkers', are inevitably biased in their assessments [5]. Accordingly, our results, based on a database of articles previously marked as 'reliable' and 'unreliable', are objective only to the extent that understanding of fake news by the database creators allows.

We can envision several possible ways to continue and develop this research. First, we could conduct an analysis of specific news publications, comparing the focus of media devoted to different subjects or having different audiences. Secondly, there is a possibility of comparing data across countries, cities, or diverse types of localities that might contribute to a better understanding of the fake news phenomenon. In addition, there is another possible option of the study development – that is the analysis not only of the headlines, but also of the main texts of articles. This method can be combined with the sentiment analysis of the news, which will provide more detailed and higher-quality information about the phenomenon being studied.

Acknowledgments. The article was prepared within the framework of the HSE University Basic Research Program.

References

1. Allcott, H., Gentzkow, M.: Social media and fake news in the 2016 election. J. Econ. Perspect. **31**(2), 211–236 (2017)
2. Basov, N.: The ambivalence of cultural homophily: field positions, semantic similarities, and social network ties in creative collectives. Poetics (78), 101353 (2020). https://doi.org/10.1016/j.poetic.2019.02.004
3. Liu, B.: Sentiment analysis: Mining opinions, sentiments, and emotions. Cambridge University Press, Cambridge (2015). https://doi.org/10.1017/CBO9781139084789
4. Bryanov, K., Vziatysheva, V.: Determinants of individuals' belief in fake news: a scoping review determinants of belief in fake news. PLoS ONE **16**(6), e0253717 (2021)
5. Ceron, W., Sanseverino, G.G., de-Lima-Santos, M.F., et al.: COVID-19 fake news diffusion across Latin America. Soc. Network Anal. Min. **11**(1), 1–47 (2021). https://doi.org/10.1007/s13278-021-00753-z
6. Cha, Y., Cho, J.: Social-network analysis using topic models. In: Proceedings of the 35th International ACM SIGIR Conference on Research and Development in Information Retrieval, pp. 565–574 August 2012

7. Karell, D., Freedman, M.: Sociocultural mechanisms of conflict: combining topic and stochastic actor-oriented models in an analysis of Afghanistan, 1979–2001. Poetics (78), 101403 (2020). https://doi.org/10.1016/j.poetic.2019.101403

8. Koirala, A.: COVID-19 fake news dataset. Mendeley Data (2021). https://doi.org/10.17632/zwfdmp5syg.1. Accessed 26 Sept 2021

9. Dant, T.: Knowledge, ideology & discourse: a sociological perspective. Routledge, 158 – 160 (2013)

10. Del Vicario, M., Bessi, A., Zollo, F., et al.: The spreading of misinformation online. Proc. Nat. Acad. Sci. **113**(3), 554–559 (2016)

11. Douglas, K.M.: COVID-19 conspiracy theories. Group Process. Intergroup Relat. **24**(2), 270–275 (2021)

12. Edson, C.T., Wei Lim, Z., Ling, R.: Defining fake news. Digital Journalism (2017)

13. Nielsen, F.A.: Evaluation of a word list for sentiment analysis in microblogs. Preprint at https://arxiv.org/abs/1103.2903 (2011)

14. Trew, T.: 'What the papers say': linguistic variation and ideological difference. In: Language and Control, pp. 117–156. Routledge, London (2018)

15. Qc, I.F.: COVID-19: fear, quackery, false representations and the law. Int. J. Law Psychiatry, **72** (2020)

16. Guo, L., et al.: Coverage of the Iraq War in the United States, Mainland China, Taiwan, and Poland: A transnational net-work agenda-setting study. J. Stud. **16**(3), 343–362 (2015)

17. Guo, L., McCombs, M.: Network Agenda Setting: The Third Level of Media Effects. Paper presented at the ICA, Boston (2011)

18. Guo, L., Vargo, C.: The power of message networks: a big-data analysis of the network agenda setting model and issue ownership. Mass Commun. Soc. **18**(5), 557–576 (2015)

19. Holland, P.W., Leinhardt, S.: An exponential family of probability distributions for directed graphs. J. Am. Stat. Assoc. **76**(373), 33–65 (1981)

20. Hunter, D.R., Handcock, M.S.: Inference in curved exponential family models for networks. J. Comput. Graph. Stat. **15**(3), 565–583 (2006)

21. Hunter, D.R., Handcock, M.S., Butts, C.T. et al.: ergm: A package to fit, simulate and diagnose exponential-family models for networks. J. Stat. Softw. **24**(3), nihpa54860 (2008)

22. Lippmann, W.: Public opinion. Routledge (2017)

23. McCombs, M.E., Shaw, D.L.: The agenda-setting function of mass media. Public Opin. Q. **36**(2), 176–187 (1972)

24. R Development Core Team: a language and environment for statistical computing. R Foundation for Statistical Computing, Vienna, Austria. ISBN 3–900051–07–0, Version 2.6.1 (2007)

25. Robinson, D., Misra, K., Silge, J.: widyr: Widen, Process, then Re-Tidy Data. https://CRAN.R-project.org/package=widyr (2021). Accessed 30 Sept 2021

26. Robins, G., Pattison, P., Kalish, Y., Lusher, D.: An introduction to exponential random graph (p∗) models for social networks. Soc. Netw. **29**(2), 173–191 (2007)

27. Tandoc, E.C., Jr., Lim, Z.W., Ling, R.: Defining 'fake news': a typology of scholarly definitions. Digit. J. **6**(2), 137–153 (2018)

28. Pedersen, T.L.: ggraph: An Implementation of Grammar of Graphics for Graphs and Networks. https://CRAN.R-project.org/package=ggraph (2021). Accessed 30 Sept 2021

Mapping Crypto: Analysis of Cryptocurrency Controversies Based on Latour's Approach

Kseniia Alikova[✉]

BDC Consulting, Moscow, Russia
writetoxen@gmail.com

Abstract. Crypto market is characterized by high volatility, an abundance of fraud and asymmetric information. Despite the declared advantages: decentralisation, independence, security and transparency, cryptocurrency is perhaps one of the most striking manifestations of "risk society". The discussion about cryptocurrencies still has a variety of debated aspects and a plurality of positions. However, the proponents of all these positions, while speaking theoretically, do not disclose the motivation of the actual actors of this market – the users of cryptocurrencies. Based on Latour's idea of "following the actors", we map the crypto community controversies and trace their connections with discourse, technologies and other actors in the cryptocurrency field, using the InfraNodus tool to analyze the crypto discussion on Twitter.

Keywords: Cryptocurrency · Controversies mapping · Social media analysis

1 Introduction

"To the Moon!" is an often-used phrase by crypto-enthusiasts on Twitter. Tabloids are full of news about the rapid rise of new altcoins. The excitement around cryptocurrencies is fueled by the growth of Bitcoin. In January 2022, its price was over $35,000, a dramatic increase from costing just $1 ten years ago. Studies proved that public sentiment has a greater impact on the price of cryptocurrencies and vice versa [1, 2, 7]. Comparing the number of term "cryptocurrency" searches on Google and the price of Bitcoin makes this correlation obvious (see Figs. 1 and 2).

This market is characterized by high volatility, an abundance of fraud, and asymmetric information [2, 4]. Despite the declared advantages: decentralisation, independence, security and transparency, cryptocurrencies are perhaps one of the most striking manifestations of "risk society".

Such ambiguity is reflected on all levels of cryptocurrency discourse. Different states make almost opposite decisions about the future of cryptocurrencies. In 2021 alone, El Salvador approved a bill that would recognize bitcoin as an official means of payment in the country, while China decided to ban any activity related to cryptocurrencies.

A. Antonyuk and N. Basov (Eds.): NetGloW 2022, LNNS 663, pp. 36–53, 2023.
https://doi.org/10.1007/978-3-031-29408-2_3

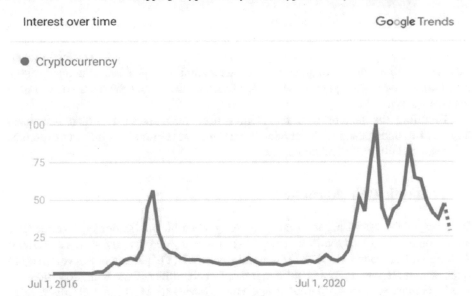

Fig. 1. Relative worldwide interest to the topic of "cryptocurrency" based on web searches on Google (June 2016–June 2022). Source: Google Trends[1]

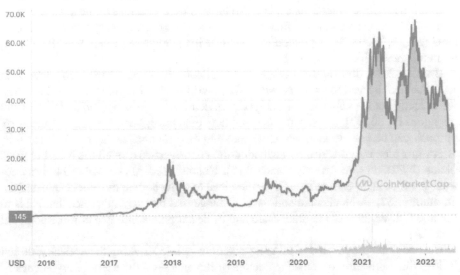

Fig. 2. The Bitcoin price (June 2016–June 2022). Source: CoinMarketCap[2].

---

[1] https://trends.google.com/trends/explore?date=all&q=cryptocurrency.

[2] https://coinmarketcap.com/currencies/bitcoin/.

These controversies make it difficult to make administrative and business decisions regarding cryptocurrencies. Moreover, the study of the cryptocurrency market is hampered not only by the disagreements described above, but also by its essential properties: decentralisation and fragmentation of actors and their positions, the huge role of non-human actors (computer code), lack of information about the actors due to their anonymity, etc.

Therefore, the main problem for researchers of this market is that there is no clear approach to understanding the cryptocurrency market or a method that can overcome the described difficulties of studying it.

2   Theoretical Background

When cryptocurrencies began to gain popularity, scientific publications sparked debates about their social and political role. Cryptocurrency enthusiasts such as Andreas Antonopoulos, Brian Kelly, David Golumbia, Nathaniel Popper and others voiced opinions, which sceptics such as Paul Krugman, Robert Shiller, Dirk Baur, Jon Baldwin, David Golumbia, Joel Z. Garrod, Nigel Dodd, Bartoletti Massimo and others vehemently opposed. The discussion about cryptocurrencies to this day is characterized by a plurality of positions. In particular, there is still no consensus even on whether a "cryptocurrency market" exists at all.

Such ambiguity can be observed at all levels. Scientists not only have many conflicting opinions about cryptocurrency, but also engage in discussions about a variety of its aspects. They focus on issues of finance [8, 12, 15, 31], government regulation and legal status [13, 17, 36, 38, 39], cryptocurrencies' role in the modern economy [3, 10, 11, 16] and management [6, 10, 18, 19, 26].

Back in 2018, Jon Baldwin conducted a Bitcoin discourse analysis to show what views are held by various well-known authors writing about cryptocurrencies [4]. The idea of using Latour's approach to analyze cryptocurrency was proposed by Molling and colleagues [5, 30, 32]. These authors show how the cartography method and the ANT approach can be used to facilitate understanding of this complex topic. Frederico Barros put this idea into practice by analyzing the controversies around the Non-Fungible Tokens (NFTs) cryptocurrency standard [5]. Najmul used ANT to describe disagreements in blockchain communities that often lead to splits both in the blockchain and the community [32]. A blockchain split is any change in a blockchain's rules that leads to permanent divergence of the blockchain and its development into two or more potential paths [34].

However, the proponents of all these positions, while speaking theoretically, do not disclose the motivation of the actual actors of this market – the users of cryptocurrencies, ultimately making up this field and being its driving force. Therefore, researchers need pay attention to the crypto discourse of real market participants and compare it to academic discourse about cryptocurrencies.

3 Method

Based on Latour's idea of "following the actors" we map the crypto community controversies and trace their connections with discourse, technologies and other actors in the cryptocurrency field [21].

Using ANT [22] as the theoretical foundation, we may address emerging issues in the sociological analysis of the cryptocurrency field. There are several other reasons why Latour's approach is productive in this field:

- Controversy mapping is a technique intended for observing and describing social debates on technology issues. This approach helps present different positions, topics and authors in the discussion [21, 37].
- ANT does not make an a priori distinction between human and non-human actors [23]. This is especially important in a field in which computer code is an actor, both discursively [20] and technically. Fundamentally, the actions and interactions of human actors in this sphere are regulated by the blockchain code and the technical ecosystem of cryptocurrencies.
- ANT does not make an a priori distinction between micro and macro actors. It allows us to analyze the actors in the network on different levels [24]. Due to the decentralization, anonymity and multiplicity of interaction forms between actors, it is difficult to determine the scale of the participants. Famous authors get into discussions with the readership, big companies and individuals conduct identical operations, etc.
- Latour's approach provides the analysis of technology adoption processes: its development, implementations, and assessments [25]. Crypto adoption is one of the most controversial issues on this topic.

In the field of cryptocurrencies, the crypto community and the agents' actions mainly exist in a digital embodiment, such as the use of online services, interaction through digital systems and representative acts (like posts and comments) on the Internet. Thus, using cartography is quite effective, since a significant part of all of the above is recorded on the Internet, which can be used to map this field more easily than unrecorded actions on offline markets.

The crypto market is developing in a state of constant self-reflection. In the words of Anthony Giddens: "the social practices are constantly examined and re-formed in the light of incoming information about those very practices, thus constitutively altering their character". Studying the dynamics of the cryptocurrency discourse allows us to trace the process of the spontaneous development of this society "in vivo" and see how innovative practices are incorporated into our lives and become the "black box".

3.1 Data Collection and Analysis

Data from Twitter posts was collected through its API. Most popular posts related to cryptocurrencies were collected daily from July 2021 to September 2021. The most popular posts were selected based on the number of reactions to them. To track the dynamics of the main topics and authors of crypto discourse on Twitter, repeated measurements were carried out after several months.

The analysis was carried out using the InfraNodus tool [35]. As a result of applying the tool, we get a directed network where nodes are normalized words (lemmas) and edges are their co-occurrences. Once a text is represented as a network in this way, a wide range of network analysis tools is then applied to detect communities of closely related concepts or topical clusters, identify the most influential nodes (top keywords), conduct sentiment analysis etc. The main themes were identified according to the Louvain community identification algorithm [9]. It detects words that are more commonly used together than others and puts them in one group, indicated by a separate color. It then ranks them by the number of each node's connections, helping to highlight the main topics. The most influential nodes are either the ones with the highest betweenness centrality or the ones with the highest degree. Sentiment analysis is used in cartography because it is not enough just to register the topics discussed, but it is also important to understand the tone of the analyzed discourse for different communities. The sentiment analysis is based on a fixed AFINN dictionary [33]. The method and algorithm of data processing using this tool are described in more detail in an article by Dmitry Paranyushkin, the creator of the tool [35]. In addition, we create a network of authors participating in the controversies where links indicate that they have mentioned each other in their posts.

The positions of scientists regarding cryptocurrencies were analyzed using the cartography of titles and abstracts of 995 articles. The sample was collected from articles with the term "cryptocurrency" in the title, indexed in Google Scholar and published between 2011 and 2021. The analysis of collected data was conducted in the same way as described above using the InfraNodus tool [35].

4   Findings

4.1   There is No Homogenous Community of Cryptocurrency Users

Although the well-established term "cryptocurrency community" is often used by researchers and media, there is no homogenous community of cryptocurrency users – the communities are separate.

The authors of the cryptocurrency discourse do not tend to unite into a single community over time, with no new edges between communities being formed. On the contrary, new clusters have appeared, and edges within the communities have been destroyed. Even the key authors have changed (see Figs. 3, 4 and 5).

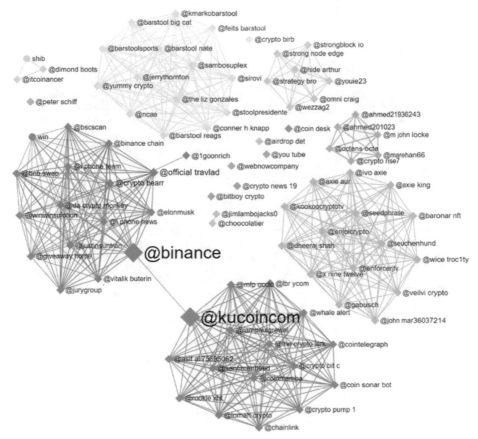

Fig. 3. Network of Twitter user accounts mentioned in relation to the term "cryptocurrency" in July 2021. The account names that are more commonly used together are indicated by a separate color. The size of the nodes indicates their betweenness centralities.

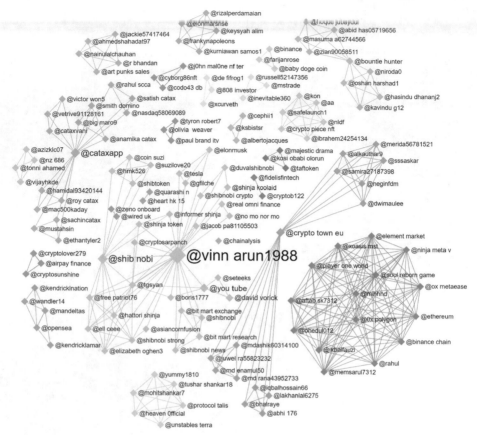

Fig. 4. Network of Twitter user accounts mentioned in relation to the term "cryptocurrency" in January 2022. The account names that are more commonly used together are indicated by a separate color. The size of the nodes indicates their betweenness centralities.

Fig. 5. Network of Twitter user accounts mentioned in relation to the term "cryptocurrency" in May 2022. The account names that are more commonly used together are indicated by a separate color. The size of the nodes indicates their betweenness centralities.

4.2 Crypto Discourse is More Positive on Twitter Than in Academic Research

The authors of scientific papers about cryptocurrency use a more negative tone than the authors of Twitter posts on this topic (see Figs. 6 and 7). On the one hand, this may be due to the fact that market participants are more loyal to cryptocurrencies, and therefore they use and write about them. On the other hand, this may be due to the placement of advertisements that present information about cryptocurrencies on social networks in a positive way in order to attract more customers. Cryptocurrency brands actively educate new people about cryptocurrencies, promote crypto publications with positive connotations and thus act as a driving force behind crypto adoption.

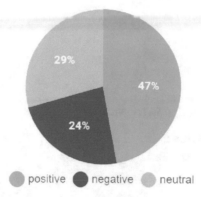

Fig. 6. Sentiment analysis of scientific papers about cryptocurrency published in 2010–2021

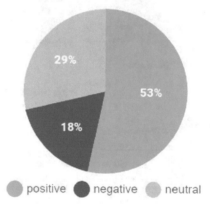

Fig. 7. Sentiment analysis of most popular posts on Twitter about cryptocurrency published in July–September 2021

4.3 Scientific Discourse is More Focused on Technology and Risks of Crypto, While Twitter Posts Are More About Earning and Top Brands

The mapping of academic research on cryptocurrency demonstrated its complexity and the lack of consensus on several issues. The main concerns are the definition of "digital money", the operational capacities of cryptocurrencies without a financial institution, their impacts on the economy, the problematics of trust and its future development (see Fig. 8).

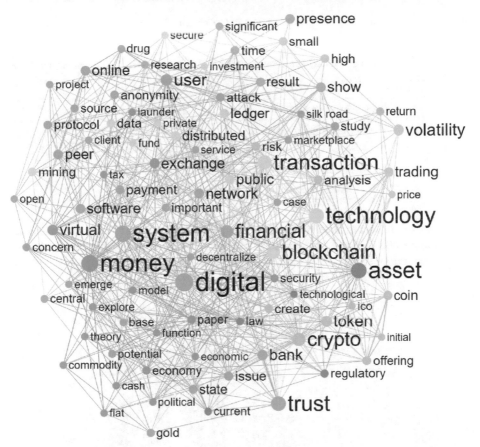

Fig. 8. Network of word co-occurrences in titles and abstracts of 995 scientific papers about cryptocurrency published in 2010–2021. The words that are more commonly used together are indicated by a separate color. The size of the nodes indicates their betweenness centralities.

The scientific literature also focuses on the discussion of technical and financial aspects of the development of cryptocurrencies as well as the risks. The words "digital", "money", "system", "asset" are the most important elements of the discussion – they have the highest measure of betweenness centrality (see Table 1).

Table 1. Discourse statistics of scientific papers about cryptocurrency published in 2010–2021

Node name	Degree	Frequency	Betweenness
digital	44	75	0.171772
money	50	81	0.147690
system	52	64	0.144186
asset	40	44	0.125004

Other notable words for academic discourse about cryptocurrencies are "blockchain", "transaction", "technology", "issue", "attack", etc. These words are included in the main topical groups of the network and are also mainly associated with technology (see Table 2).

Table 2. Main topical groups in the discourse of scientific papers about cryptocurrency published in 2010–2021

Keywords in main topical groups	Percentage of group
system, digital, financial	17%
money, white paper, exchange	14%
blockchain, transaction, technology	13%
issue, user, attack	12%

However, in discussions on Twitter, the question of whether cryptocurrencies are considered to be money is not raised at all. There are also practically no discussions about the risks and technical problems of the blockchain (see Fig. 9).

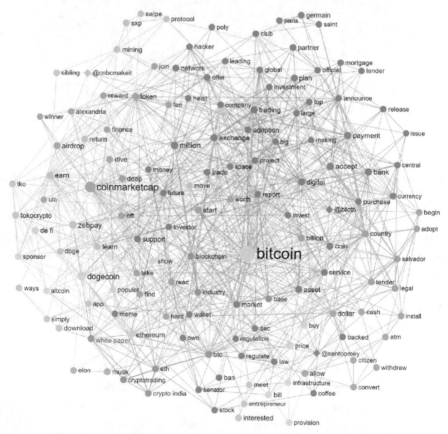

Fig. 9. Network of word co-occurrences in most popular Twitter posts about cryptocurrency published in July–September 2021. The words that are more commonly used together are indicated by a separate color. The size of the nodes indicates their betweenness centralities.

Research shows that the key themes are related to brand names "Bitcoin", "CoinMarketCap", "Dogecoin" (see Table 3) and also to the words "earn", "exchange", "million", which emphasizes the role of economic motivation in this field (see Table 4).

Table 3. Discourse statistics of the most popular Twitter posts about cryptocurrency published in July–September 2021

Node name	Degree	Frequency	Betweenness
Bitcoin	129	476	0.797129
CoinMarketCap	37	39	0.214590
Dogecoin	27	123	0.092908

Table 4. Main topical groups in the discourse of the most popular Twitter posts about cryptocurrency published in July–September 2021

Keywords in main topical groups	Percentage of group
exchange, million, digital currency	36%
earn, learn, zebpay	13%
CoinMarcetCap, token, NFT	12%
payment, announce, accept	9%

The most discussed topics in this market are related to finance: "exchange", "million", "digital currency", "earn". Despite the fact that cryptocurrencies are based on technology and ideology of blockchain, people are more interested in the issue of money.

4.4 The Growth of the NFT Market is Related to Its Popularity

It is important to stress that the cryptocurrency discourse is incredibly dynamic, so the most discussed topics often change and new influential concepts and authors emerge. A very indicative example is the growth in popularity of NFTs. In 2021, few people knew about this type of token. In the summer of 2021, NFT ranked 14th in terms of frequency of mention and had a degree of 28. For comparison, Bitcoin, the traditional leader of all posts about cryptocurrencies, had a degree of 131 (see Fig. 10 and Table 5).

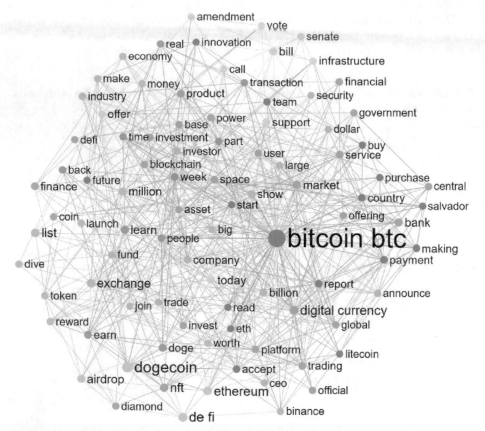

Fig. 10. Network of word co-occurrences in the most popular posts on Twitter related to the term "cryptocurrency" published in June 2021. The words that are more commonly used together are indicated by a separate color. The size of the nodes indicates their betweenness centralities.

Table 5. Main topical groups in the discourse of the most popular Twitter posts about cryptocurrency published in June 2021

Keywords in main topical groups	Percentage of group
exchange, million, trade	21%
bank, trading, market	21%
bitcoin btc, payment, report	21%
list, learn, earn	16%

In January 2022, in the network of mentions of the term "cryptocurrency" on Twitter, the term NFT bypassed even the first cryptocurrency – its topical cluster ranked first (see Fig. 11 and Table 6).

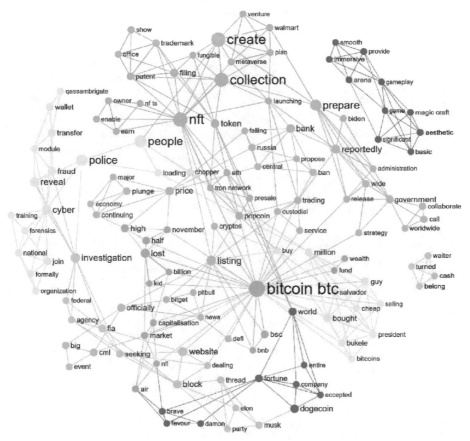

Fig. 11. Network of word co-occurrences in the most popular posts on Twitter related to the term "cryptocurrency" published in January 2022. Words that are more commonly used together are indicated by a separate color. Node size indicates their betweenness centralities.

Table 6. Main topical groups in the discourse of the most popular Twitter posts about cryptocurrency published in January 2022

Keywords in main topical groups	Percentage of group
nft, create, collection	16%
president, people, million	12%
bitcoin btc, lost, capitalisation	10%
cyber, national, transfer	10%

In May 2022 NFT, Bitcoin and Ethereum topic groups merged (see Table 7). NFT still maintained its influence, becoming a more significant element in the topic of Bitcoin and Ethereum discussion (see Fig. 12).

Fig. 12. Network of word co-occurrences in the most popular posts on Twitter related to the term "cryptocurrency" published in May 2022. Words that are more commonly used together are indicated by a separate color. Node size indicates their betweenness centralities.

Table 7. Main topical groups in the discourse of the most popular Twitter posts about cryptocurrency published in May 2022

Keywords in main topical groups	Percentage of group
nft, eth, bitcoin	19%
list, web, bsv	17%
market, thebriden	12%
early, exchange	12%

Therefore, in these networks we can trace over time the dynamics of the increase in the discussion of NFTs in the general discourse around cryptocurrencies. As with the correlation between Bitcoin price growth and search queries, NFTs are gaining popularity and rising in price. An active discussion of this topic leads to an increase in the interest of actors. And they, in turn, buy NFTs and launch new projects, which boost the popularity of this type of token even more (see Fig. 13).

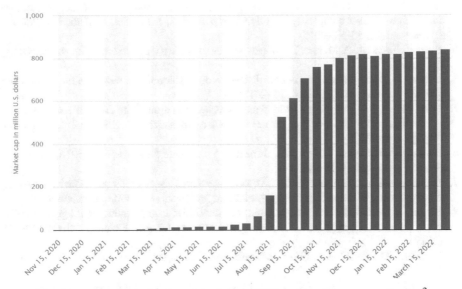

Fig. 13. Market capitalization of Art Blocks NFT projects. Data source: Statista[3]

5 Conclusion

Exploring the debate about cryptocurrencies through cartography opens up great prospects for researchers.

This approach avoids many problems associated with the ambiguity of this field and allows us to take advantage of the benefits of digital network studies. Cartography of the cryptocurrency market based on Latour's approach allows us to analyze the structure and dynamics of the discourse of this rapidly changing market. From a theoretical point of view, ANT is also a productive approach to analyzing this market. The mass adoption of innovations, endowing computer code with subjectivity, and interaction between actors of different levels in the same field – all these features are inherent in this market, and therefore Latour's approach seems to be the most relevant for further studies on cryptocurrencies.

This study shows that using cartography of controversies in academic and Twitter discourse, one can analyze the structure and dynamics of the communication network, the differences in topics and their connotations, as presented by a variety of authors, the connection between popular themes, the demand for technical solutions in this market, and much more.

---

[3] https://www.statista.com/statistics/1291885/art-blocks-nft-market-cap/.

References

1. Aloosh, A., Ouzan, S.: The psychology of cryptocurrency prices. Fin. Res. Lett. **33**, 101192 (2020). https://doi.org/10.1016/j.frl.2019.05.010
2. Antonakakis, N., Chatziantoniou, I., Gabauer, D.: Cryptocurrency market contagion: market uncertainty, market complexity, and dynamic portfolios. J. Int. Fin. Markets, Inst. Money, Elsevier, 61(C), 37–51 (2019)
3. Aste, T., Tasca, P., Di Matteo, T.: Blockchain technologies: the foreseeable impact on society and industry (2017)
4. Baldwin, J.: In digital we trust: bitcoin discourse, digital currencies, and decentralized network fetishism. Palgrave Commun. **4**, 14 (2018)
5. Barros Frederico. Dropping Art: mapping controversies around the 2021 NFT Craze. Interactive film and media conference: new narratives, racialization, global crises, and social engagement (2021)
6. Bheemaiah, K.: Why business schools need to teach about the blockchain (2015). https://dx.doi.org/10.2139/ssrn.2596465
7. Bianchi, D., Dickerson, A., Babiak, M.: Trading volume and liquidity provision in cryptocurrency markets. Working Paper series (2022). https://dx.doi.org/10.2139/ssrn.3239670
8. Bleher, J., Dimpfl, T.: Today I got a million, tomorrow, I don't know: on the predictability of cryptocurrencies by means of Google search volume. Int. Rev. Financ. Anal. **63**, 147–159 (2019)
9. Blondel, V.D., et al.: Fast unfolding of communities in large networks. J. Stat. Mech.: Theory Exp. (2008)
10. Catalini, C., Joshua, S.G.: Some simple economics of the blockchain. Commun. ACM **63**(7), 80–90 (2020)
11. Cong, L.W., Li, X., Tang, K., Yang, Y.: Crypto wash trading. arXiv preprint arXiv: 2108.10984 (2021)
12. Davidson, S., De Filippi, P., Potts, J.: Disrupting governance: the new institutional economics of distributed ledger technology (2016). https://dx.doi.org/10.2139/ssrn.2811995
13. Gandal, N., Halaburda, H.: Can we predict the winner in a market with network effects? Competition in Cryptocurrency Market. Games, 7 (June 27, 2016)
14. Hacker, P., Thomale, C.: Crypto-securities regulation: ICOs, token sales and cryptocurrencies under EU financial law. Euro. Company Fin. Law Rev. **15**(4), 645–696 (2018)
15. Herring, S.C.: Computer-mediated discourse analysis: An approach to researching online behavior. In: Barab, S., Kling, R., Gray, J.H. (eds.) Designing for Virtual Communities in the Service of Learning, pp. 338–376. Cambridge University Press, Cambridge (2004)
16. Hileman, G., Rauchs, M.: 2017 Global Blockchain Benchmarking Study. Cambridge Centre for Alternative Finance (2017). https://dx.doi.org/10.2139/ssrn.3040224
17. Huberman, G., Leshno, J.D., Moallemi, C.: Monopoly without a monopolist: an economic analysis of the bitcoin payment system. Rev. Econ. Stud. **88**(6), 3011–3040 (2021)
18. Hughes, S., Middlebrook, S.: Regulating cryptocurrencies in the United States: current issues and future directions (2014)
19. Karajovic, M., Kim, H.M., Laskowski, M.: Thinking outside the block: projected phases of blockchain integration in the accounting industry. Aust. Account. Rev. **29**(2), 319–330 (2019)
20. Kazan, E., Tan, C.W., Lim, E.T.: Value creation in cryptocurrency networks: towards a taxonomy of digital business models for bitcoin companies (2015)
21. Knorr Cetina K., Bruegger U.: The market as an object of attachment: exploring postsocial relations in financial markets. Canadian J. Soc. **25**(2), 141–168 (2000)
22. Latour, B.: Networks, societies, spheres: reflections of an actor–network theorist. Int. J. Commun. **5**, 796–810 (2011)

23. Latour, B.: Science in Action: How to Follow Scientists and Engineers Through Society. Harvard University Press, Cambridge, MA (1987)
24. Latour, B.: The impact of science studies on political philosophy. Sci. Technol. Hum. Values **16**(1), 3–19 (1991)
25. Latour, B., Callon, M.: Unscrewing the big Leviathan: how actors macro-structure reality and how sociologists help them to do so. In: Knorr-Cetina, K., Cicourel, A.V. (eds.) Advances in Social Theory and Methodology, pp. 277–303. Routledge, London, England (2014)
26. Latour, B.: Science in action: how to follow scientists and engineers through society (1987)
27. Lee, J. Y.A decentralized token economy: how blockchain and cryptocurrency can revolutionize business. Business Horizons 62.6 (2019)
28. Li, X., Wang, C.A.: The technology and economic determinants of cryptocurrency exchange rates: the case of Bitcoin. Decis. Support Syst. **95**, 49–60 (2017)
29. Marres, N.: Why Map Issues? On controversy analysis as a digital method. Sci. Technol. Hum. Values **40**(5), 655–686 (2015)
30. Marres, N., Moats, D.: Mapping controversies with social media: the case for symmetry **1**(2) (2015)
31. Molling, G., Klein, A., Hoppen, N., et al.: Cryptocurrency: a mine of controversies. J. Inf. Syst. Technol. Manage. **17**, (2020)
32. Morisse, M.: Cryptocurrencies and bitcoin: charting the research landscape. AMCIS (2015)
33. Najmul Islam, A.K.M., Mäntymäki, M., Turunen, M.: Why do blockchains split? An actor-network perspective on Bitcoin splits, Technological Forecasting and Social Change, vol. 148, (2019)
34. Nielsen, F.: A new ANEW: Evaluation of a word list for sentiment analysis in microblogs. arXiv preprint arXiv:1103.2903 (2011)
35. Nyman, L., Mikkonen, T., Lindman, J., Fougere, M.: Perspective of code forking and sustainability in open source software. In: Hammouda, I., Lundell, B., Mikkonen, T., Scacchi, W., (Eds.), Open Source Systems: Long-term Sustainability, Proceedings of the 8th IFIP WG 2.3 International Conference, OSS, vol. 378, pp. 274–279 (2012)
36. Paranyushkin, D.: InfraNodus: generating insight using text network analysis. In: The World Wide Web Conference (WWW '19). Association for Computing Machinery, New York, NY, USA, pp. 3584–3589 (2019)
37. Riley, J.: The current status of cryptocurrency regulation in China and its effect around the world. China WTO Rev. **7**(1), 135–152 (2021)
38. Venturini, T.: Diving in magma: how to explore controversies with actor-network theory. Public Understanding Sci. **19**(3), 258–273 (2009)
39. Werbach, K.: Trust, but verify: why the blockchain needs the law. Berkeley Technol. Law J. **33**(2), 487–550 (2018)
40. Wright, A., De Filippi, P.: Decentralized blockchain technology and the rise of lex cryptographia (2015)

Spatially Embedded Networks

Mapping Daily Consumption of Urban Community Services Based on Public Participation GIS Data

Oleksandra Nenko[1](✉) and Anastasia Galaktionova[2]

[1] Institute for Advanced Studies, University of Turku, Turku, Finland
oleksandra.nenko@utu.fi
[2] ITMO University, St. Petersburg, Russia

Abstract. In the paper we construct and describe everyday activity network (EAN) as a representation of city structure emerging in the context of daily consumption. The dataset used for EAN construction is based on a sociological georeferenced survey conducted with a public participation geoinformation system. The survey covers 1938 citizens of the resort city of Sochi, Russia, who have mapped community services they regularly use and their homeplaces. The EAN unites households and community facilities as nodes through the links of visitations and was constructed from ego-networks of the households and the visited facilities. The georeferenced data allows to project the EAN and its separate clusters on a map and explore the geographical features in terms of urban planning. Through a community detection procedure 16 communities were allocated, 7 of the largest were considered in terms of geographical and social factors, which influence their network and spatial configuration. The detected communities can be considered as neighborhoods and show specifics of urban planning in Sochi. In the conclusions we consider how an EAN of a city can be used for drawing recommendations for planning the infrastructure of community facilities.

Keywords: PPGIS · Activity spaces · Everyday activity network

1 Introduction

The turbulence of the current times has dramatically transformed the world's cities whose populations grow exponentially. The consumption of community services, such as shops, parks, cultural venues, remains a regular practice of urban inhabitants, and growth of the urban population leads to the growth of demand for these services. If the spatial distribution of the services does not meet the demand and does not correlate with the everyday practices in space it may increase the inequality in terms of service accessibility. The facilities where different services are provided form a physical layer of the urban environment, which has been studied carefully through the lens of quality of life and well-being, quite often at the intersection of geographic, economic and sociological methods.

© The Author(s), under exclusive license to Springer Nature Switzerland AG 2023
A. Antonyuk and N. Basov (Eds.): NetGloW 2022, LNNS 663, pp. 57–67, 2023.
https://doi.org/10.1007/978-3-031-29408-2_4

In this paper we construct an everyday activity network of the city to show an approach to address the spatial patterns of everyday life. Though the network analysis approach to study urban structure has a long tradition, its applications addressing city space usage in the context of consumption are still scarce. To do so we use data obtained through a public participation geographic information system (PPGIS) survey, which allows us to link the households of the respondents with the facilities they use on a map.

The paper's structure is as follows. First, a literature review of approaches to analyze the geography of everyday life is introduced. Then the data set and data gathering method are described. After that the results are provided and interpreted. Finally, we explain how the EAN concept might be applied in planning practice.

2 Literature Review

There are two major approaches to analyze the geography of the everyday life of the city residents: the spatial analysis of activity spaces and the network analysis of behavioral patterns in space. The former one focuses on measuring characteristics of discrete activity spaces of individuals; the latter gives opportunity to consider interconnected individuals and places they use.

The activity space approach has been used to define a set of places personally visited since the early 1970s [1, 2]. Also, other terms described similar concepts: action spaces [3], home range [4], territorial range [5] and neighborhood [6]. There are four methods for allocating the activity space [2, 4, 7]: the deviational ellipse, the minimum convex polygon, the shortest network path, and kernel density estimation. These measures allow to build the borders of an activity space based on a set of destinations visited by an individual. However, all four of them have a basic disadvantage: they construct an abstract space, the whole of which in reality might not be used and interacted with excluding the places visited [2].

The network analysis approach allows analyzing data at three different levels: the micro-level of an individual, the meso-level of a community, and the macro-level of a network as a whole. A fundamental advantage is that exploration of the topological characteristics of relationships between objects is possible rather than the geometrical ones. There is a tendency to apply network analysis to a wide variety of network nodes, including countries, cities, bus stops, firms, artists, travelers, social media users. However, a meta-analysis of papers in human geography starting from 1990 suggests that research is still much more attracted to exploring the node level of networks and their measures (e.g. network density or centralization), while the importance of ties is still understudied [8]. In the area of human geography, network analysis of connections, such as the small-world method, has only recently started being applied in papers [9–12]. Examples cover detecting communities with optimized measures of modularity [13–15] based on the social and spatial proximity of the individuals belonging to them. It is important to highlight that analysis of social patterns of space usage on the city scale is mostly based on analysis of big transactional data, such as mobile or bank transactions [16]. Yet the study based on knowledge of micro interactions of individuals with urban infrastructure, e.g. community facilities, during their everyday life activities is underdeveloped.

This paper aims to fill in this gap by considering the patterns of everyday activities on a city-scale as an everyday activity network (EAN) of the city. We construct an EAN based on the sociological georeferenced data on personal visitations of community facilities described below. This kind of data reflects the micro-level of actors' interaction with space during consumption practices. Additionally, the network methods allow a transition from the micro-level (the ego-networks) to the macro-level (the joint network of households-facilities). By constructing and studying an EAN of the city, detecting communities in it and projecting them on a map, we show the heuristic potential of studying the topology in the patterns of everyday space usage with network tools.

3 Study Area and Data

3.1 Study Area

The study area for the paper is the resort city of Sochi, located in the southern part of Russia at the Black Sea. Its population is 720 thousand people and the density of population is 2450 people/km^2. Spreading along the sea coast and delimited by the Caucasus mountains from the shore, Sochi is officially the longest city in Russia and the second longest city in the world (the length of Sochi is 145–148 km). From the urban planning view, there are four urban centers in Sochi (created by merging former villages) rounded by a series of small seaside resorts, between which there are vast underurbanized territories. The city of Sochi underwent considerable development during the Winter Olympic Games of 2014, when a new urbanized subcentre of Krasnaya Polyana appeared and when urban infrastructure of Sochi for inhabitants (cultural, sports, public venues) also grew. In 2021 the city administration commissioned a new master plan, which was the framework for the survey described below.

3.2 Data and Survey Method

The data considered in the paper was obtained from a sociological georeferenced survey conducted with a public participation geographic information system "Mapsurvey". This PPGIS was developed by a team including the authors of the paper. It combines possibilities of online mapping of respondents' evaluations and spatial practices with a traditional sociological questionnaire. The survey was conducted within a research framework of studies anticipating the design of a new master plan for Sochi led by the Institute of Perspective Urban Development[1]. The aim of the survey was to receive information about recreational practices of city inhabitants and public venues they regularly visit. The link to the online survey was distributed through the city authorities and local communities, so the survey gained massive public attention. The sample considered in this paper contains 1938 people who have marked the venues they use as well as the locations of their homes on a map. 55% of respondents were women, 45% were men, which corresponds to the statistical gender distribution. 84% of participants were between the ages of 26 and 55, thus covering the active working-age population.

---

[1] http://niipgrad.spb.ru/.

The participants drew 5353 markers of places they visit regularly. The survey design allowed respondents to map places representing community facilities: health care, education, sports, entertainment, and open public spaces, such as parks and promenades. Besides mapping the places they visit, the participants were asked to answer which kinds of community facilities should be increased in number in Sochi.

3.3 Analysis of Data

The construction of the EAN went as follows.

The dataset contained venues of two types – homeplaces and community facilities, which were considered as two modes of network nodes. The EAN was constructed in a two-step approach. First, the ego-networks of households and the respective community facilities they use were constructed (representing each respondent of a survey considered here). Then a joined network was constructed by merging the ego-networks of households.

To dig further we assume that the EAN cannot be a homogeneous network and it should reveal certain clustering tendencies, while usage of the city space is also uneven and is shaped by social and spatial factors. To check this assumption, we have applied the community detection algorithm to the EAN and then checked for significance of the dependent variables available in the dataset.

Before the construction of the networks, procedures to reduce data noise were undertaken. The initial markers of community facilities, which were mapped by the respondents, were repetitive and abusive. To allocate the set of unique facilities, we had to convert the markers into places. We used a heuristic approach and created buffers of 50 m around all markers. Afterwards we merged the buffers of the markers, which had an identical attribute – kind of facility. 50 m size was chosen as optimal after comparing results from 10 m up to 100 m; buffers less than 50 m divided some spacious facilities, such as parks, into several areas, while buffers of more than 50 m size, on the contrary, united clearly different facilities (e.g., a park and an embankment). This procedure resulted in unique 509 places representing 14 kinds of facilities; the kinds were coded in a more detailed way than the initial number of kinds given in the survey based on the environmental data about the location of different facilities. This aggregated number of places contained almost a half of the initial markers (2427) (Fig. 1 A). All the cartographical analysis was run in open software QGIS 3.22.7 [17].

After reducing the data, we created the EAN in two steps. Network analysis and visualization was run in Gephi 0.9.5 [18]. Firstly, an aggregated set of places were connected with the households of participants to create ego-networks (Fig. 1 B). Secondly, we have joined the ego-networks into a unified network of homeplaces and community facilities at the city scale – the everyday activity network of Sochi (EAN). The resulting EAN contains 1436 nodes and 2331 links with a set of attributes – socio-demographic characteristics of the respondents and typology of the facilities. To detect clusters of everyday activities in the EAN we run a community detection algorithm in Gephi 0.9.5. To check the statistical significance of different attributes of nodes we have run a non-parametric Kruskal-Wallis test, as the distribution of the parameters in the communities is not normal.

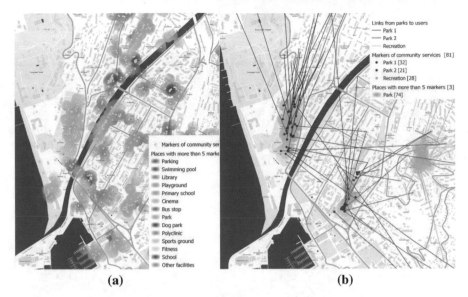

Fig. 1. A. The map of 509 places. **B.** The example of ties between places and homes

4 Results

The EAN of Sochi has a rather low connectivity (1.62) and a low density (0.2%). After running the community detection algorithm, 81 communities with modularity level 0.81 were detected. To filter out less important connections we have chosen for further analysis 16 communities, which contain 92% of all links of the EAN. The structure of the EAN is subject to the core-periphery law which gains its specific appearance in urban space. The largest communities detected are located in the central region of Sochi, the smallest communities with less connected nodes tend to the peripheral areas of the city.

The specifics of 16 clusters in terms of spatial characteristics, social-demographic features of their users and kinds of facilities in them, are apparent in the results of the non-parametric Kruskal-Wallis test. The relationship is statistically significant for distances in-between the homeplaces and facilities, respondent's age and kind of the community facility (Table 1).

Table 1. Kruskal-Wallis test for statistical significance of dependent variables in 16 communities

	χ^2	df	p	ε^2
Median length of tie, m	126.4	15	**< .001**	0.0589 < 0.2
Total length of ties, km	116.6	15	**< .001**	0.0543 < 0.2
Employment	27.0	15	0.029	0.0126 < 0.2
Age	46.2	15	**< .001**	0.0215 < 0.2

(continued)

Table 1. (*continued*)

	χ^2	df	p	ε^2
Gender	33.7	15	0.004	$0.0157 < 0.2$
Kind of facility	155.5	15	**< .001**	$0.0725 < 0.2$

To look closer into the clusters we have chosen 7 largest of them, each containing more than 5% of network elements (Fig. 2).

Fig. 2. The 7 largest communities in the Sochi EAN

The statistical relations with dependent variables in these clusters are similar as for all of the clusters (Table 2).

Table 2. Kruskal-Wallis test for statistical significance of dependent variables in the 7 largest communities

	χ^2	df	p	ε^2
Median length of tie, m	34.4	6	**< .001**	0.0240
Total length of ties, km	35.9	6	**< .001**	0.0250
Employment	20.7	6	0.002	0.0144
Age	28.2	6	**< .001**	0.0196

(*continued*)

Table 2. (*continued*)

	χ^2	df	p	ε^2
Gender	13.2	6	0.040	0.0092
Kind of facility	52.3	6	**< .001**	0.0364

The chosen 7 communities aggregate 193 places marked by 648 respondents (Fig. 3). The geographical location and spatial proximity influence the belonging of a node, representing a facility or a household, to the community. For example, the largest community (#45, green) mostly covers places at the seaside in the central part of Sochi; the second largest community (#57, magenta) includes places along Sochi river; the community #14 (violet blue) covers places at the seaside in Adler district in the periphery of Sochi; the community #70 (violet) includes places in the northern part of central Sochi. It is worth noting that there is no community formed by the venues of the sports and leisure infrastructure built during the Winter Olympic Games; this means that this infrastructure is not at all used by Sochi inhabitants in their daily life.

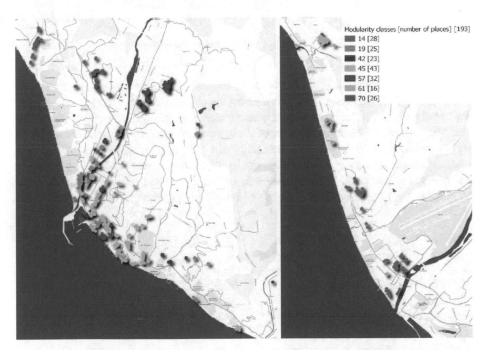

Modularity classes [number of places] [193]
- 14 [28]
- 19 [25]
- 42 [23]
- 45 [43]
- 57 [32]
- 61 [16]
- 70 [26]

Fig. 3. The map of seven largest communities in the Sochi EAN

Aside from the geographical factors shaping clusters of the EAN, there are other factors making some of the detected communities overlap in space. This might be explained by multifunctionality of places in them and social-demographic features of users. Further consideration of the detected communities makes evident that most of the links in

64   O. Nenko and A. Galaktionova

them are formed by connections of homeplaces to the open public spaces (mostly parks), schools and polyclinics. The distribution of the links by kind of facilities in each of the 7 communities is illustrated below (Fig. 4). The only community without polyclinic (#19) includes a popular swimming pool and a cinema. It is interesting how availability of specific facilities in the geographical proximity forms the community (e.g. visitations of fitness clubs form part of the connections in communities #19, 45, 57).

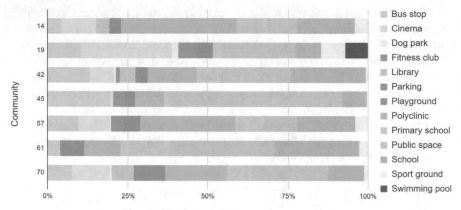

Fig. 4. Number of ties in seven communities

The socio-demographic characteristics of the communities show that these densely connected EAN clusters are mostly formed by the patterns of space usage particular for employed people and freelancers in their active working age – 26–55 years (Fig. 5). This might be interpreted as an inclination of these working age groups to actively use the city space and visit the facilities in their daily life. People of younger age, the students, are forming spatial connections in community #19, which does not have a polyclinic, but has a swimming pool. Younger age groups are less present in community #61, where the participants of elderly age (55 years and over) form most of the spatial connections.

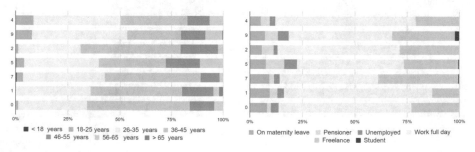

Fig. 5. Age and employment of participants in seven communities

5 Discussion

It should be noted that this empirical study has limitations that need to be addressed in a future design. The present study is conducted on a dataset collected through a PPGIS method. Online mapping surveys may result in some increased spatial and thematic biases in relation to other methods of measurement such as mobile phone data and GPS tracking. To obtain the opinion of younger and older age groups, it is necessary to conduct additional research using a field survey. Due to the specifics of the low level of participation in the discussion of public life (among younger groups) and due to the low level of digital literacy (in the case of older age groups) online participatory mapping does not always collect the opinions of these groups.

Some of the measures we employed rely on the number of locations marked on a map. This number may be limited while respondents' engagement in mapping activity and their level of mapping skills might not represent the full account of their everyday practices. However, the errors pertain to individual participants, and there is no known reason for errors to be systematically related to the variables of interest of the study. Therefore, we can safely assume that the individual errors do not introduce strong biases to the aggregate analysis presented in the paper.

6 Conclusions

An EAN can be considered as a representation of a city through patterns of everyday usage which actually connect homeplaces with community facilities. Its measures may reveal the character of everyday consumption of community services, which may be loosely connected if the activities are weak. The decomposition of the network into communities (modularity classes) helps identifying neighborhoods. The consideration of these neighborhoods may show stable patterns in consumption practices of inhabitants who belong to them, as well as unique features which are provoked both by the availability of certain infrastructure and readiness to use it. These considerations might be thought-provoking for urban planners to build a more effective urban infrastructure system supplying the demand.

Communities detected in the studied EAN reveal the real structure of the Sochi residential area, different from the administrative division. The revealed community structure shows a clear subdivision into different areas that separate the residents' activity space into neighborhoods. The results suggest that some separations were caused by geography, for example, elongated planning structure of Sochi and its historical structure consisting of subcenters, as well as the existence at the seaside where many facilities gravitate. Some partitions are caused by social factors, such as age, employment and consequent demand for certain kinds of facilities. At the same time there are certain enduring consumption patterns; schools, polyclinics and parks pertain to each community and may indicate healthy neighborhoods.

Further EAN research can help renovation projects of built-up city districts by allowing more spatially flexible decisions without disrupting existing spatial patterns of consumption. The development of new community facilities inside existing EAN clusters can be considered as an alternative to ideas of normative accessibility of 300–800 m [19] or a 15-min city [20]. The study of an EAN might also help in localizing the positions for new facilities. For example, in Sochi the gravitation of clusters to the coast gives the authorities a hint to consider the placement of requested facilities within the boundaries of the revealed communities at the seaside.

Acknowledgements. This paper was supported in 2021 by Russian Science Foundation, project 21-77-10098 "Spatial segregation of the largest post-Soviet cities: analysis of the geography of personal activity of residents based on big data".

References

1. Horton, F.E., Reynolds, D.R.: Effects of urban spatial structure on individual behavior. Econ. Geogr. **47**(1), 36–48 (1971)
2. Patterson, Z., Farber, S.: Potential path areas and activity spaces in application: a review. Transp. Rev. **35**(6), 679–700 (2015)
3. Dijst, M.: Action space as planning concept in spatial planning. Netherlands J. Housing Built Environ. **14**(2), 163–182 (1999)
4. Hasanzadeh, K., Broberg, A., Kyttä, M.: Where is my neighborhood? A dynamic individual-based definition of home ranges and implementation of multiple evaluation criteria. Appl. Geogr. **84**, 1–10 (2017)
5. Broberg, A., Kyttä, M., Fagerholm, N.: Child-friendly urban structures: Bullerby revisited. J. Environ. Psychol. **35**, 110–120 (2013)
6. Kyttä, A.M., Broberg, A.K., Kahila, M.H.: Urban environment and children's active lifestyle: SoftGIS revealing children's behavioral patterns and meaningful places. Am. J. Health Promot. **26**(5), e137–e148 (2012)
7. Perchoux, C., Chaix, B., Brondeel, R., Kestens, Y.: Residential buffer, perceived neighborhood, and individual activity space: new refinements in the definition of exposure areas–The RECORD Cohort Study. Health Place **40**, 116–122 (2016)
8. Glückler, J., Panitz, R.: Unleashing the potential of relational research: a meta-analysis of network studies in human geography. Prog. Hum. Geogr. **45**(6), 1531–1557 (2021)
9. Etxabe, I., Valdaliso, J.M.: Measuring structural social capital in a cluster policy network: insights from the Basque Country. Eur. Plan. Stud. **24**(5), 884–903 (2016)
10. Glückler, J., Panitz, R.: Relational upgrading in global value networks. J. Econ. Geogr. **16**(6), 1161–1185 (2016)
11. Henderson, W.D., Alderson, A.S.: The changing economic geography of large US law firms. J. Econ. Geogr. **16**(6), 1235–1257 (2016)
12. Prota, L.: Toward a Polanyian network analysis: market and non-market forms of coordination in the rice economy of Vietnam. J. Econ. Geogr. **16**(6), 1135–1160 (2016)
13. Sobolevsky, S., Szell, M., Campari, R., Couronne, T., Smoreda, Z., Ratti, C.: Delineating geographical regions with networks of human interactions in an extensive set of countries. PLoS ONE **8**(12), e81707 (2013)
14. Cranshaw, J., Schwartz, R., Hong, J.I., Sadeh, N.: The livehoods project: utilizing social media to understand the dynamics of a city. In: International AAAI Conference on Weblogs and Social Media (2012)
15. Nenko, A., Koniukhov, A., Petrova, M.: Areas of habitation in the city: improving urban management based on check-in data and mental mapping. Commun. Comput. Inf. Sci. **947**, 235–248 (2019)

16. Ratti, C., et al.: Redrawing the map of Great Britain from a network of human interactions. PLoS ONE 5(12), e14248 (2010)
17. QGIS 3.22.7 (2022). https://www.qgis.org/en/site/forusers
18. Bastian, M., Heymann, S., Jacomy, M.: Gephi: an open source software for exploring and manipulating networks. In: International AAAI Conference on Weblogs and Social Media (2009)
19. Local standards for urban design of Sochi city. https://sochi.ru/upload/iblock/583/583fffb2d c424cdf5908a65e8ddbd993/ad431ab783bdf05a13e24c420210d3ad.pdf
20. Moreno, C., Allam, Z., Chabaud, D., Gall, C., Pratlong, F.: Introducing the "15-Minute City": sustainability, resilience and place identity in future post-pandemic cities. Smart Cities 4(1), 93–111 (2021)

A Scalable Spatio-Temporal Analytics Framework for Urban Networks

Yuri Bogomolov[1,3] and Stanislav Sobolevsky[1,2,3(✉)]

[1] Masaryk University, Brno, Czech Republic
ss9872@nyu.edu
[2] New York University, New York, USA
[3] ITMO University, St. Petersburg, Russia

Abstract. Numerous real-world processes and events, especially in the urban domain, are represented with spatio-temporal transactional data (STTD): examples include human and vehicle mobility, communication, and economic transactions. Despite the overwhelming variability of such data sets they can be seen as having very much in common, including their structure as well as the analytic and visualization challenges they face. The present paper describes an idea of an analytic platform implementing an underlying general data model as well as the analytic and modeling tools for STTD. The platform is intended to provide increased scalability of the STTD-driven applications enabling broad reuse of the most common analytic and visualization solutions in multiple contexts of urban analytics.

Keywords: Spatio-temporal transactional data · Spatio-temporal transactional networks · Urban networks · Analytics framework

1 Introduction and Literature Review

Since our life happens in space and time, the majority of data sets describing real-world processes possess a spatio-temporal component, containing information about space and time of recorded events. The revolutionary growth of spatio-temporal data attracts increasing attention of researchers and practitioners [12, 13, 16, 27] offering analytic and visualization platforms [7, 10, 11, 14, 25, 29, 33].

However, often spatio-temporal data describing urban activities possess an additional important component. Since city as a complex system consists of multiple types of interacting actors (for example people, businesses, vehicles, and buildings), many events described by urban data represent interactions (transactions) between pairs or sets of objects happening in space and time: e.g. people interacting with each other (phone calls, meetings, social connections), with businesses (customer transactions, employment) as well as with urban infrastructure and transportation. This way one can call such data "spatio-temporal transactional data" (STTD).

Different types of STTD have been proven to serve as a useful proxy for human mobility, including geotagged social media [8, 18, 21, 22, 24], cell phone

© The Author(s), under exclusive license to Springer Nature Switzerland AG 2023
A. Antonyuk and N. Basov (Eds.): NetGloW 2022, LNNS 663, pp. 68–78, 2023.
https://doi.org/10.1007/978-3-031-29408-2_5

data records [6,20], credit card transactions [31] or Bluetooth sensor readings [34]. But as none of the data sets alone use to provide a comprehensive proxy for urban mobility, there seems to be a clear benefit in bringing multiple data layers together enabling multi-layered analytic solutions [8,35]. Such data were also successfully leveraged for several urban applications, such as regional delineation [26,32], land use classification [23] or transportation optimization [28], as well as for research applications such as developing and evaluating the models of human mobility or interaction [17].

This concept paper presents an idea and a proof-of-concept implementation of a multi-layered STTD framework. The framework will allow saving multiple efforts to implement similar sorts of analytic and visualization solutions of comparable functionality in different contexts, which otherwise require custom implementation from scratch. Consequently, this can accelerate the development of STTD solutions and applications, increasing their scalability.

2 Materials

2.1 STTD Data Model

The STTD structure could be represented as a set of actor objects and a set of binary transactions between pairs of objects happening in space and time. Please note that two sets described above form a spatio-temporal transactional network (STTN). The input dataset may contain multiple transactions between the same pair of actors, and some transactions have an asymmetric nature: a commute from A to B is different from a commute from B to A. Based on these requirements we get the directed multigraph model:

$$G = (V, A, s, t) \tag{1}$$

V – a set of nodes (actor objects)

A – a set of arcs (transactions between objects)

Directed edges (arcs) have source and target nodes: if an arc a goes from v_1 to v_2 then v_1 is called the source node of a and v_2 is the target node of a. In the definition (1) this information is represented by two functions:

$s : A \mapsto V$ – a source function of an arc

$t : A \mapsto V$ – a target function of an arc

The model (1) is sufficient to represent the network structure, for example, we can define a network for landline phone call data: every user is represented by a node, while every arc represents a phone call (transaction). Function s returns the source (caller), while function t returns the destination (callee) of the call. Yet, in addition to the network structure, data analytics and modeling problems often require extra knowledge about actors (e.g. geo-coordinates) and transactions (examples include call duration or start time). Node-level data can be incorporated into node labels, while arc labels can store transaction-level data. Therefore, we get a directed multigraph with labeled nodes and arcs:

$$G_L = (V, A, \Sigma_V, \Sigma_A, s, t, L_V, L_A) \tag{2}$$

There are four new components compared to the directed graph definition (1):

Σ_V, Σ_A – alphabets of node and arc labels, that define the information we store about nodes and transactions

$L_V : V \mapsto \Sigma_V$ – a node labeling function

$L_A : A \mapsto \Sigma_A$ – an arc labeling function

In the phone call example, we needed to know the geo-coordinates of landline phones, then Σ_V represents all possible geo-coordinates, while L_V returns the coordinates for a given user. If we want to know the start and end time of a phone call then Σ_A may be represented by $(start\_time, end\_time)$ – a set of two-element vectors, and L_A returns vectors for a given call.

Node and arc labels can contain vectors of data, including multiple time and space components, but visualization and modeling methods need to know what time and space are used for a given research problem. As result, STTN can be formally defined as a directed multigraph with labeled nodes and arcs, and with space and time labels:

$$STTN = (V, A, \Sigma_V, \Sigma_A, s, t, L_V, L_A, space, time) \tag{3}$$

In a general case nodes and arcs may have space and time components:

$space : \Sigma_V \cup \Sigma_A \mapsto Space$ – a space function of node and arc labels. *Space* is usually represented as a polygon, a list of polygons, or a single point in space.

$time : \Sigma_V \cup \Sigma_A \mapsto Time$ – a time function of Σ_A or Σ_V

Both functions can return an empty element if a given dimension is not defined. In the case of phone calls only users have spatial coordinates, while the time dimension is defined only for phone calls.

In contrast to the well-defined time series, space series, and graph models, which cover only one dimension of data, the proposed model (3) combines the spatial, temporal, and network components within one coherent representation. The examples in Sect. 2.3 and network categories in Sect. 2.2 demonstrate the wide diversity of urban datasets that can be presented as an STTN.

2.2   STTN Categories

STTN nodes (actors) could be static, dynamic, or abstract (non-spatial). Static objects are characterized by a fixed geospatial location (e.g. businesses, taxi zones, locations themselves), while dynamic objects do not have a fixed location (e.g. people, vehicles) and their current locations can be identified through local transactions with static objects (customer visiting a store, cell phone user served by a cell tower). Abstract objects do not have a spatial dimension at all (e.g. social groups, concepts, topics). In turn, transactions are classified as local (assuming that interacting objects should share the same location), remote (no spatial constraints), or abstract (one of the objects is an abstract one, so the transaction does not have a special dimension). Also, the transactions are classified as instant (happening at a single moment of time) and lasting (the connection between objects once established lasts until one of the objects is involved in a corresponding type of transaction canceling the given one).

Based on the number of object classes STTNs are categorized as single-class (e.g. taxi rides) or two-class (e.g. credit card transactions) networks. In case an interaction involves more than two object classes at the same time it can be decomposed into several two-class STTNs (in the geo-tagged social media example below we built user-location and user-topic networks).

2.3   STTN Examples

Taxi Rides and similar modern transportation datasets are usually defined at the zone level due to data privacy concerns (for example NYC taxi zones [3]). Taxi zones can correspond to network nodes, while every ride can be represented as an arc from the origin node to the destination. From the modeling perspective, we may be interested in the geography of the rides, ride prices, and the number of passengers. The full STTN definition is provided below:

V – a set of taxi zones

A – a set of taxi rides

Σ_V – a set of all possible geo-objects

Σ_A – a set of vectors that contain start time, end time, cost, and passenger count taxi ride characteristics.

$s : A \mapsto V$ – a function that returns the origin node of the ride

$t : A \mapsto V$ – a function that returns the destination node of the ride

$L_V : V \mapsto \Sigma_V$ – a mapping between taxi zones and zone labels

$L_A : A \mapsto \Sigma_A$ – a mapping between taxi rides and taxi ride labels

$space$ – a function that returns the geo-object of the taxi zone

$time$ – a function that returns the start time of the ride (the first component of the Σ_A vector)

Please note that there are multiple ways to define STTN for the same dataset. Framework users may want to use the end time of the ride as a time dimension, or even aggregate all rides between a pair of nodes into a single arc that has relevant statistics (like the total number of rides/passengers) in the label vector. The STTN framework aims to provide a generic data model, that can be used to achieve various modeling needs.

Geo-tagged Social Media Records represent a very valuable source for human mobility studies [8,21,22]. Human mobility modeling problems usually operate with two groups of objects: people (or users) and locations. Therefore, STTN nodes can be represented by a union of user and location nodes, while a list of social media records forms a bipartite graph.

V – a union of user and location objects

A – a set of social-media posts

Σ_V – a union of geo-locations and the empty element

Σ_A – a set of timestamps

$s : A \mapsto V$ – a function that returns the author of the post

$t : A \mapsto V$ – a function that returns the location of the post

$L_V : V \mapsto \Sigma_V$ – a mapping between nodes and node labels (the mapping function returns the empty object for user nodes and a geo-object for location nodes)

$L_A : A \mapsto \Sigma_A$ – a mapping between posts and post labels

space – a function that returns the geo-object of the location node (or the empty element)

time – a function that returns the post timestamp

If we study topics or content of the social media records then we can define V as a union of users and topics. In that case, the space component can be moved to the arc labels.

Credit and Debit Card Transactions is another common dataset type that has space and time dimensions in addition to relationships between customers and points of sale. Local transactions in static points of sale represent more value for human mobility studies since they reveal information about customer location. Local payments data can be modeled with STTN in the following way:

V – a union of customers and points of sale

A – a set of transactions

Σ_V – a union of geo-locations and the empty element

Σ_A – a set of vectors that contain the time and amount components

$s : A \mapsto V$ – a function that returns the customer who initiated the transaction

$t : A \mapsto V$ – a function that returns the point of sale for the transaction

$L_V : V \mapsto \Sigma_V$ – a mapping between nodes and node labels (the mapping function returns the empty object for customer nodes and a geo-object for point of sale nodes)

$L_A : A \mapsto \Sigma_A$ – a mapping between transactions and transaction properties

space – a function that returns a geo-object (or the empty element for customer nodes)

time – a function that returns the transaction timestamp (the first component of the Σ_A vector)

2.4   Transformations, Analytics and Modeling API

Data Retrieval is a tedious, complex, and error-prone process: even simple datasets have multi-page specification documents that describe various edge cases. The need to join multiple datasets contributes to even higher complexity. STTN provides a unified data representation, that allows implementing the data retrieval logic once per dataset, and sharing the result retrieval functionality with the community.

Aggregation of source data is commonly used in data analysis to get multiple datasets to the same granularity level or to increase data coverage per compute unit. STTN supports two types of aggregations: node-level (aggregate census blocks to census tracts, or to city districts) and transaction-level (aggregate all rides between the same origin and destination nodes). The graph structure of STTN helps the framework to automate some aggregation steps: for example, when the nodes are aggregated (taxi zones to city districts), origin and destination nodes in all transactions are automatically updated to appropriate city districts.

Modeling of human mobility or transportation needs uses an STTN-compatible data model, which allowed us to provide a generic implementation of gravity and radiation models. Of special interest are graph neural network models [19], including the novel augmentation hierarchical graph neural networks [30]. The list of models can be extended to support any model that takes a time/space series, a network or a combination of these structures.

Community Detection is a popular technique in urban data analysis. Common applications include dimensionality reduction for big urban datasets and city delineation. Identified communities' nodes are frequently clustered into a single node, which in turn requires updates to the list of transactions. STTN allows automation of these steps and provides a simple interface that outputs a new STTN, where detected communities represent nodes.

Anomaly and Pattern Detection methods are well-studied independently for time series or network datasets. However, these methods consider only one dimension of spatio-temporal networks and therefore have limited detection power. The unified data model creates an opportunity to define new anomaly and pattern detection techniques that leverage all data dimensions and can be applied to a wide variety of STTN inputs.

2.5   Engineering Requirements and Prototype Design

Urban data processing combines elements of time, space, and network analysis. Due to the large scope of functional requirements, we decided to use the composition engineering principle and rely on the most popular Python framework for each corresponding area:

- Pandas for time-series functionality
- GeoPandas for spatial operations
- NetworkX for network analytics
- PyTorch for graph neural network modeling

The composition principle allows operating with an STTN as a whole object, easily converting the STTN to a Pandas/GeoPandas/NetworkX object, running required computations, and attaching the results back to the original STTN model before moving to the next steps.

The large size of modern urban datasets implies strict requirements on the framework's performance. Reused utility methods allow STTN to promote engineering best practices: for example, data is downloaded in batches to reduce memory footprint and the result files are cached on disk using Apache Parquet [1] data format, which reduces the file size and processing time up to ten times for real datasets [5]. The proof-of-concept implementation of the framework [4] was used to process millions of datapoints. One of the applications is described in the section below.

2.6   Framework Application

We used the framework prototype described above to study human mobility patterns across US cities in our previous work [9]. We provide a detailed description of the analysis steps below to demonstrate how STTN helped us to automate and scale various data processing and modeling steps.

Understanding the home-work commute is an important challenge for transportation planning. The main objective of the paper was to study the impact of income on the home-work commute. Our work was based on the Origin-Destination Employment Statistics Census dataset, published by the Longitudinal Employer-Household Dynamics (LEHD) program [2]. The dataset forms a network: every record is represented as an edge between the origin (home) and destination (work) nodes. Therefore, the STTN can be constructed in the following way:

V – a set of census blocks
A – a set of origin-destination commute pairs
Σ_V – a set of geo-regions
Σ_A – a set of vectors that contain the year, low and high-income commute flow values
$s : A \mapsto V$ – a function that returns the origin block of the commute pair
$t : A \mapsto V$ – a function that returns the destination block of the commute
$L_V : V \mapsto \Sigma_V$ – a mapping between nodes and node labels
$L_A : A \mapsto \Sigma_A$ – a mapping between transactions and transaction properties
space – a function that returns a geo-region of a census block
time – a function that returns data timestamp (census year)

The gravity model [36] is a popular tool to predict commute flows between network nodes. The model is similar to Newton's theory of gravity, and it assumes that the number of trips from location i to j is proportional to the 'attraction' values of origin and destination nodes, while activity tends to fall with distance. The gravity model is specified as:

$$T_{i,j} = \frac{O_i D_j}{f(r_{i,j})} \tag{4}$$

In (4) $T_{i,j}$ represents the number of trips from i to j, O_i and D_j are 'attraction' values and $f(r_{i,j})$ is a distance deterrence function. In the case of home-work commute networks, origin attraction can be represented by node population P_i, while destination attraction is usually measured by the number of jobs J_j. The power law function is the most popular choice for the distance deterrence function. As a result, the model looks like this:

$$T_{i,j} = k \frac{P_i J_j}{r_{i,j}^q} \tag{5}$$

where k is a normalizing coefficient, and q is a model parameter that needs to be fitted. Please note that the exponent q represents commuters' sensitivity to trip distance.

The LEHD census dataset provides a breakdown of commuters by income level, which allowed us to build two networks for each city: one for the low-income population and another one for the high-income population. The source data provides census block-level data for the entire state, but we needed zip code-level data for major US cities. STTN filtering and aggregation functionality proved to be very helpful: during node filtering, the framework leverages the node index to efficiently filter out all irrelevant transactions, while the node aggregation functionality automatically recomputes geo-coordinates, based on the coordinates of the input nodes. Then we used arc aggregation to combine data about all commuters between a given zip code pair.

After network construction, the model fitting step requires P_i, J_j, and $r_{i,j}$ for all pairs of network nodes. Node population (P_i) can be computed as a sum of commute values in outgoing arcs of node i, while the number of jobs (J_j) can be computed as a sum of commute values in incoming arcs of node j. Both of these computations are supported by the aggregation STTN API, described above.

To fit the model we take *log* of (5) and get

$$log(T_{i,j}) = log(k) + log(P_i) + log(J_j) - q\,log(r_{i,j}) \tag{6}$$

q in (6) can be fitted using a linear regression model, where $log(k)$ represents the intercept. The fitted q value measures the heterogeneity in the distance sensitivity of commuter groups.

We trained gravity model coefficients for the 12 largest US cities (Fig. 1) and found that the exponents for low-income groups are always more than exponents for high-income groups, except Los Angeles (which is known to be the least affordable city in the US [15]). In particular, it means that the high-income population tends to have a slower mobility decay with increasing distance compared to the low-income population.

The unified data model and the STTN framework helped us to scale up the data processing and analysis steps to the 12 largest US cities, or 24 networks, some of which had more than one million edges. The data retrieval layer allowed us to build a network for any US city with one line of code, while the aggregation and modeling API helped to implement the gravity model fitting logic for a generic STTN. As result, the standardized STTN data model not only helped us to scale the analysis to 24 networks but also provided a way to build reusable

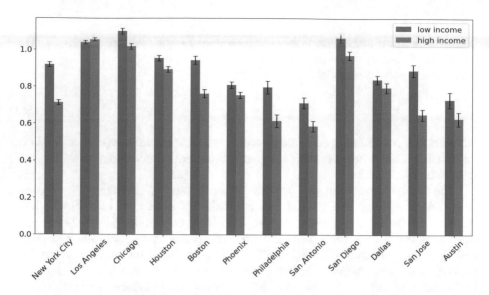

Fig. 1. Gravity model exponents for high and low-income groups with 95% confidence intervals across 12 largest US cities. The figure demonstrates the framework application from our previous work [9].

analysis blocks, including LEHD data retrieval and gravity model fitting, that we already started using in new studies.

3   Conclusion

In this work, we proposed a generic data model that can serve as a baseline for developing analytic, modeling, and visualization functionality for spatio-temporal transactional data. We implemented a limited version of the framework that supports single-type static objects and applied it to scale up our commute analysis to cover the twelve largest US cities [9].

We plan to expand the STTN framework in three directions:

- Support new datasets and dataset types
- Add more analytical, modeling, and visualization functionality
- Develop custom analysis techniques for STTD (for example, an anomaly detection method that leverages space and time dimensions as well as network structure)

The framework helps to automate data retrieval, filtering, aggregation, pattern and anomaly detection, visualization, and modeling steps. At the same time, the unified data model allows the introduction of novel analysis methods for a wide variety of spatio-temporal transactional datasets.

Acknowledgements. This research was partially supported by the Masaryk University Award in Science and Humanities and the RSF grant 21-77-10098 "Spatial segregation of the largest post-Soviet cities in the Russian Federation: analysis of the geography of personal activity of residents based on big data".

References

1. Apache Parquet. https://parquet.apache.org/. Accessed 31 May 2022
2. Longitudinal employer-household dynamics. https://lehd.ces.census.gov/data/. Accessed 31 May 2022
3. NYC taxi zones. https://data.cityofnewyork.us/Transportation/NYC-Taxi-Zones/d3c5-ddgc. Accessed 29 May 2022
4. STTN implementation. https://github.com/yuribogomolov/sttn. Accessed 31 May 2022
5. What is Apache Parquet. https://databricks.com/glossary/what-is-parquet. Accessed 31 May 2022
6. Amini, A., Kung, K., Kang, C., Sobolevsky, S., Ratti, C.: The impact of social segregation on human mobility in developing and industrialized regions. EPJ Data Sci. **3**(1), 1–20 (2014). https://doi.org/10.1140/epjds31
7. Andrienko, N., Andrienko, G.: A visual analytics framework for spatio-temporal analysis and modelling. Data Mining Knowl. Disc. **27**(1), 55–83 (2013)
8. Belyi, A., et al.: Global multi-layer network of human mobility. Int. J. Geographical Inf. Sci. **31**(7), 1381–1402 (2017)
9. Bogomolov, Y., He, M., Khulbe, D., Sobolevsky, S.: Impact of income on urban commute across major cities in US. Procedia Comput. Sci. **193**, 325–332 (2021)
10. Cao, G., Wang, S., Hwang, M., Padmanabhan, A., Zhang, Z., Soltani, K.: A scalable framework for spatiotemporal analysis of location-based social media data. Comput. Environ. Urban Syst. **51**, 70–82 (2015)
11. Compieta, P., Di Martino, S., Bertolotto, M., Ferrucci, F., Kechadi, T.: Exploratory spatio-temporal data mining and visualization. J. Vis. Languages Comput. **18**(3), 255–279 (2007)
12. Cressie, N., Wikle, C.K.: Statistics for Spatio-Temporal Data. Wiley, Hoboken (2015)
13. Diggle, P.J.: Statistical Analysis of Spatial and Spatio-Temporal Point Patterns. CRC Press, New York (2013)
14. Ferreira, N., Poco, J., Vo, H.T., Freire, J., Silva, C.T.: Visual exploration of big spatio-temporal urban data: a study of New York City taxi trips. IEEE Trans. Visual Comput. Graphics **19**(12), 2149–2158 (2013)
15. Flaming, D., et al.: Los Angeles rising: a city that works for everyone (2015)
16. Gao, S.: Spatio-temporal analytics for exploring human mobility patterns and urban dynamics in the mobile age. Spatial Cogn. Comput. **15**(2), 86–114 (2015)
17. Grauwin, S., et al.: Identifying and modeling the structural discontinuities of human interactions. Sci. Rep. **7**(1), 1–11 (2017)
18. Hawelka, B., Sitko, I., Beinat, E., Sobolevsky, S., Kazakopoulos, P., Ratti, C.: Geo-located Twitter as proxy for global mobility patterns. Cartogr. Geogr. Inf. Sci. **41**(3), 260–271 (2014)
19. Kipf, T.N., Welling, M.: Semi-supervised classification with graph convolutional networks. arXiv preprint arXiv:1609.02907 (2016)

20. Kung, K.S., Greco, K., Sobolevsky, S., Ratti, C.: Exploring universal patterns in human home-work commuting from mobile phone data. PLoS ONE **9**(6), e96,180 (2014)

21. Kurkcu, A., Ozbay, K., Morgul, E.: Evaluating the usability of geo-located Twitter as a tool for human activity and mobility patterns: a case study for NYC. In: Transportation Research Board's 95th Annual Meeting, pp. 1–20 (2016)

22. Paldino, S., Bojic, I., Sobolevsky, S., Ratti, C., González, M.C.: Urban magnetism through the lens of geo-tagged photography. EPJ Data Sci. **4**(1), 1–17 (2015). https://doi.org/10.1140/epjds/s13688-015-0043-3

23. Pei, T., Sobolevsky, S., Ratti, C., Shaw, S.L., Li, T., Zhou, C.: A new insight into land use classification based on aggregated mobile phone data. Int. J. Geogr. Inf. Sci. **28**(9), 1988–2007 (2014)

24. Qian, C., et al.: Geo-tagged social media data as a proxy for urban mobility. In: Hoffman, M. (ed.) AHFE 2017. AISC, vol. 610, pp. 29–40. Springer, Cham (2018). https://doi.org/10.1007/978-3-319-60747-4_4

25. Ratti, C., Claudel, M.: Live Singapore! The urban data collider. Transfers **4**(3), 117–121 (2014)

26. Ratti, C., et al.: Redrawing the map of Great Britain from a network of human interactions. PLoS ONE **5**(12), e14,248 (2010)

27. Roddick, J.F., Spiliopoulou, M.: A bibliography of temporal, spatial and spatio-temporal data mining research. ACM SIGKDD Explorations Newsl **1**(1), 34–38 (1999)

28. Santi, P., Resta, G., Szell, M., Sobolevsky, S., Strogatz, S.H., Ratti, C.: Quantifying the benefits of vehicle pooling with shareability networks. Proc. Natl. Acad. Sci. **111**(37), 13290–13294 (2014)

29. Senn, O., Khairul, M., Maitan, M., Pribadi, R., Shah, M., Sivaprakasam, R.: Data-collider: an interface for exploring large spatio-temporal data sets. In: SIGGRAPH Asia 2015 Visualization in High Performance Computing, pp. 1–4 (2015)

30. Sobolevsky, S.: Hierarchical graph neural networks. arXiv preprint arXiv:2105.03388 (2021)

31. Sobolevsky, S., Sitko, I., Tachet des Combes, R., Hawelka, B., Murillo Arias, J., Ratti, C.: Cities through the prism of people's spending behavior. PLoS ONE **11**(2), e0146,291 (2016)

32. Sobolevsky, S., Szell, M., Campari, R., Couronné, T., Smoreda, Z., Ratti, C.: Delineating geographical regions with networks of human interactions in an extensive set of countries. PLoS ONE **8**(12), e81,707 (2013)

33. Van de Weghe, N., De Roo, B., Qiang, Y., Versichele, M., Neutens, T., De Maeyer, P.: The continuous spatio-temporal model (CSTM) as an exhaustive framework for multi-scale spatio-temporal analysis. Int. J. Geogr. Inf. Sci. **28**(5), 1047–1060 (2014)

34. Yoshimura, Y., et al.: An analysis of visitors' behavior in the Louvre museum: a study using Bluetooth data. Environ. Plann. B. Plann. Des. **41**(6), 1113–1131 (2014)

35. Zhu, E., Khan, M., Kats, P., Bamne, S.S., Sobolevsky, S.: Digital urban sensing: a multi-layered approach. arXiv preprint arXiv:1809.01280 (2018)

36. Zipf, G.K.: The p 1 p 2/d hypothesis: on the intercity movement of persons. Am. Sociol. Rev. **11**(6), 677–686 (1946)

Ivan Turgenev's Austrian Networks

Larisa Poluboyarinova[(✉)] [iD], Olga Kulishkina[iD], and Andrei Zhukov[iD]

St. Petersburg State University, St. Petersburg, Russia
{l.poluboyarinova,o.kulishkina,a.p.zhukov}@spbu.ru

Abstract. Ivan Turgenev (1818–83) enjoyed unparalleled popularity in Western Europe during the 19th century. There has been a Turgenev boom in Austria, leading to the emergence of a group of Turgenev's devotees, a phenomenon which has long been noticed by researchers. This article establishes a connection between Turgenev's followers and admirers as a network by applying the methodology of cultural transfer research and the SNA method using epistolary heritage, diaries, literary-critical periodical publications, as well as intertextual traces observed in literary texts. As revealed in this study, there were two types of networks around Turgenev in Austria before the First World War. Firstly, throughout his stays in spa towns (eight years in Baden-Baden and two longer stays in Karlovy Vary), Turgenev established a network of personal contacts with the Austrians that the writer himself operationalized for the economics of the arts and literature, using connections for his literary and cultural business. Secondly, there was a broader network of Austrian "Turgenevians" at the time, who (mostly) perceived Turgenev's work without personal contact with the Russian master. Thus, on the basis of literary texts, literary criticism, translations, oral discussions, theatrical performances, and readings of Turgenev's texts, it is possible to reconstruct the nature of this reception. Finally, the focus is shifted to the interconnectedness of individual actors (nodes) in this network, demonstrated by two clusters: Leopold von Sacher-Masoch – Wanda von Sacher-Masoch – Ferdinand von Saar and Marie von Ebner-Eschenbach – Betty Paoli – Ida Fleischl von Marxow.

Keywords: Ivan Turgenev · Literary networks · Austrian literature · Comparative Studies · Cultural transfer · Social network analysis · Spa Town

1 Introduction

Ivan Sergeevič Turgenev (1818–83), a Russian writer who spent the second half of his life in Western Europe (in Germany, England, and France), is considered to be "the most significant intellectual-literary mediator between Russia and Europe in the nineteenth century" because of his extensive and intensive contacts "with countless intellectual greats and artists of the time" [1: 165]. Numerous studies of Turgenev's life and works have considered his biography to be an example of productive mediation between East and West. In 2018, as the bicentennial of the writer's birth was approaching, this topic appeared in several academic publications [2–6].

The widespread idea of Turgenev's exceptional "mediator" (fr. "mediateur", M. Espagne) qualities can hardly hide the fact that in relation to his personality and his

© The Author(s), under exclusive license to Springer Nature Switzerland AG 2023
A. Antonyuk and N. Basov (Eds.): NetGloW 2022, LNNS 663, pp. 79–94, 2023.
https://doi.org/10.1007/978-3-031-29408-2_6

oeuvre, it is a question of two processes. By virtue of his excellently worked out networks throughout his life, Turgenev functioned both as an actor and an intermediary, mediating his own literary works, as well as books by Russian authors, and later also texts by his German and French colleagues, through his well-functioning relationships both to the West and the East [7, 8]. Therefore, the culturally relevant international transfer of Turgenev's literary substance as such (themes, motifs, narrative attitudes, discourses, and ideologemes) could hardly happen without the author's active cooperation and interest, if not "against" his will, solely due to the growing public interest in his work and, as a consequence, an increased readiness to receive it on the part of the respective target culture, which displayed a strong "oncoming traffic" (rus. "vstrečnoe dviženie", A. Veselovsky). These factors have to be considered when analyzing Turgenev's Austrian reception. As part of this project, we also examine the cultural transfer of Turgenev's works into Austria during the 19th century, focusing on the reconstruction of Austrian Turgenevians' networks.

2   Theory

2.1   *Cultural Transfer* Research

Cultural transfer research aims to "make visible" acts and examples of cultural mediation, as well as to "separate stages of their circulation" [9: 186], emphasizing, first and foremost, the intercultural life context of the intermediary's life, i.e., "a person with an intercultural life story", who "stands at the crossroads of cultures (contexts)" [10: 117]. Given the figure of Ivan Turgenev, the "great mediator", and the impact of his literary works at the crossroads of cultures, this aspect is particularly pertinent in our case.

There are two further methodological aspects of the Cultural Transfer Method that are fundamental to our research. The first one is characterised by its propensity for field research, which is a consequence of an empiricism-driven approach. This method explores the mediators' life-world orientations, as they become tangible, especially in autobiographical and intimate contexts, i.e. "diaries, letters, etc." [10: 115]. The second aspect is its interest in the "net" both as a metaphor and as a social structure. The cultural transfer "does not happen between compact or closed cultural spaces or even between national cultures, but between groups, individuals, divided institutions, media (such as journals), etc., which is called a network (réseau)" [10: 107 (see also [11] as an attempt to unify cultural transfer and network studies).

Cultural transfer research focuses primarily on the figure of the mediator and on the cultural goods to be transferred; the networks of recipients come less into focus within the framework of this paradigm. We would like to turn our attention to this otherwise understudied network of recipients. Consequently, the main focus of our research is not so much on the mediator figure as such (Ivan Turgenev, in our case) nor on the "cultural goods" to be mediated (his works), but rather on the cross-contextual network structure (Austrian Turgenevians).

2.2   Comparative Literature and Social Network Analysis

Our research is about an egocentric network of a Russian author and a network structure formed among his successors in a cultural context outside Russia – in Austria. Such a constellation is traditionally considered a comparative literature phenomenon. Traditionally, within the framework of one paradigm, it is possible to analyze only a handful of authors – between two and five – by means of a typological or genetic comparison. When attempting to form a summarized opinion about the author's reception throughout a country, however, one must consider a greater number of his contacts. As a result, one refrains from leading with canon-specific hierarchies or fixing one's attention on the meaning or on the content of the literary text. As far as this case is concerned, social network analysis is a suitable method.

To begin with, SNA does not analyse literary phenomena from within, but seeks to understand their external social context as a whole. "Relations are not the properties of individual agents, but of systems of agents" [12: 2]. In fact, SNA considers itself more of a method for identifying and displaying these social structures, rather than a theory: "Social network analysis is neither a theory nor a methodology. Rather, it is a perspective or a paradigm. It takes as its starting point the premise that social life is created primarily and most importantly by relations and the patterns they form" [13: 22].

Thus, SNA can provide a structured overview of an author's network in the context of his both direct (personal and epistolary) and indirect (mediated through his literary production) contacts across the European or national literary fields. SNA, in addition to providing an overall comprehensive view (at the node level), also focuses on ties, i.e., specific transport channels ("pipelines" [13: 12]) and content to be transported:

> Flows are relations based on exchanges or transfers between nodes. These may include relations in which resources, information or influence flow through networks. Like interactions, flow-based relations often occur within other social relations and researchers frequently assume or study their co-existence. [13: 12]

As a consequence, SNA is relevant to our problem because it allows us to observe and "measure" Turgenev's role as a skilled networker in spreading his literary works. Moreover, when the ties are not formed by the author's mediation activity or his distributors but by the public's interest in the author's work, one can speak a fortiori of the revealed receptive structure, as a system of interconnected recipients of the author's works in the Austrian literary field.

Only recently have literary studies attempted to conduct network research [11, 14–16]. Drawing on the methods demonstrated in such studies, we combine elements of cultural transfer research, comparative literature, and SNA in our research. We employ the relational principle in approaching literary networks as demonstrated, for example, by E. Thomalla et al., i.e. a procedure in which "identities, settings, attitudes, and attributes are not conceived as antecedents and considered in isolation, but are conceived as the results of collective processes" [16: 7].

3   Data and Methodology

It is the first study that analyzes the phenomenon of Ivan Turgenev's Austrian Network using literary network theory. For the design of this analysis, an in-depth-examination of the socio-cultural context of Turgenev's reception in Austria was necessary, the results of which are presented in paragraph 4. By identifying key actors as nodes and defining power and distribution lines as ties, it outlines the problem area.

Accordingly, in our study, we follow the "soft" trend of SNA, namely "on attempting to use the concepts and theorems of network theory as analytical tools for understanding < ... > societies" [17: 3]. Nevertheless, we will work with the SNA fundamentals such as social network, actor, node, and tie.

In this respect, apart from Turgenev himself, nodes are representatives of the Austrian cultural field between 1865 and 1914 who maintained personal contacts with Turgenev and/or shared his work through any medium (discussion, translation, performance, imitation, inter-textual approach). The key dates 1865 and 1914 were chosen because Turgenev's first resounding success with the Austrian public began in 1865 with the appearance of Turgenev's "Tales" in F. Bodenstedt's translation. In 1914, due to the beginning of the First World War, the whole socio-cultural atmosphere changed radically, so that, among other things, the interest in the Russian author dropped sharply. We define ties as connections between the nodes. Turgenev was personally connected by a strong tie only to one Austrian cultural figure, namely Moritz Hartmann (see Sect. 5.1), so we eliminated the distinction between weak and strong ties. The following kinds of connections are meant: personal contacts, correspondence, but also indirect connections that run through the translation, review, staging or literary imitation or plagiarism of Turgenev's work. For us, a connection is also a case of secondary imitation, for instance, when an imitator of Turgenev is imitated: for example, Wanda von Sacher-Masoch, who in her own literary work imitates her husband Leopold von Sacher-Masoch, who in turn was Turgenev's imitator.

We distinguish between two types of ties: the first type of ties represents non-literary connections, the second type of ties represents literary-related connections. The first type of relationship is easily identifiable: these are all Austrians with whom Turgenev was in personal contact or correspondence throughout his life due to extra-literary matters. The second type of relations arose as a result of professional contacts (publications, translations, and literary criticism) between Turgenev and the relevant actors, as well as their exchanges of views on aesthetics and poetics. The literary ties are formed even if an actor involved had only occasionally interacted with Turgenev. Various studies indicate that this exchange resulted in (often mutual) creative enrichment at various levels of reception. In order to establish the ties, we considered the relevant collections of letters. Moreover, the primary texts were also reviewed. Historical-critical (academic) editions of works that document the respective "contacts" between authors or shared "influences" between the works are also important sources for the reconstruction of the network.

Among the factors we consider is the direction of contacts: they can be one-sided or reciprocal. One-sided relationships are identified as those in which there is a communication interest only on the side of one actor. This type includes among others correspondence proposals that were not accepted or were interrupted. One-sided relationships also include documented examples of an interest in the work of a fellow writer that are not based on reciprocity. We define a doubly directed edge as reciprocal personal contact or correspondence between the relevant actor and Turgenev or between the actors themselves.

Our task was, as a first step, to reconstruct Turgenev's personal relations with representatives of the Austrian cultural field so that an egocentric network with Turgenev in the middle emerged. The second step involved studying Austrian actors who shared Turgenev's works but were not directly connected to him. Turgenevians who were in relation with the other Austrian Turgenevians, communicating and practically realizing (as translations, readings, epistolary reviews of Turgenev's works) their common affinity with the Russian author, were included in the network. The Austrian Turgenevians who, despite their documented interest in Turgenev (published reviews of his works, obituaries for Turgenev, imitated his style), were not connected with any other Turgenevian (or when such a connection could hardly be identified on the basis of the available sources), were also considered. However, they were not integrated into the network of Austrian Turgenevians, but stand as individual nodes.

Turgenev's contacts from the Austrian cultural field were studied through the index and commentaries of his academic collection [18, 19]. Marie von Ebner-Eschenbach's diaries [20] and her correspondence with Josephine von Knorr [21] and Ferdinand von Saar [22] allowed us to identify lesser-known actors operating as Austrian Turgenevians without contacting the Russian author directly, and we were also able to identify their connections. Saar's correspondence with the Princess of Hohenlohe was also of great importance for this purpose [23]. The digital Österreichisches biographisches Lexikon [24] provided articles that list the personal contact information for Turgenevians among themselves, which was essential for establishing connections among them.

4 Turgenev in/and Austria

4.1 Turgenev and Austrian Literature: Overview

During Turgenev's lifetime, his "Austrian disciples" received attention, which can be seen in the critical essays by Wilhelm Goldbaum [25], Karl von Thaler [26], Otto Glagau [27] and others. A German writer, Theodor Storm, in one of his letters to Turgenev of 1868, referred to Leopold von Sacher-Masoch's novella "Moon Night". It was the "offspring" or "disobedient child" of Turgenev, according to Theodor Storm's letter to him in 1868 [28: 68], a statement the Russian author vehemently denied.

A DNA test is useful when paternity is questioned or even denied. In the course of the 20th century, such tests (as a search for a genetic relation, as a representative of comparative literature puts it) were actually conducted several times. This was particularly true in Austrian and German comparative studies, when Turgenev's reception became a popular topic for doctoral dissertations, especially until the 1980s. Although Günther Wrytzens attempted to balance the topic "Turgenev in Austria" within the framework of a large article in 1982 [29], he was unable to accomplish this for it went on with the search for Turgenev's "traces" in Austrian literature [30–32]. Since the new readings of the Turgenevians, i.e. Austrian Turgenevians who in previous research were only referred to as "late Austrian realists" standing in the shadow of the Viennese School, have generated major revaluation and reinterpretation, this search has gained even greater importance. Having previously been regarded as simply precursors of the European realism, these authors are now recognized as founders and mediators of Austrian literary modernity [33, 34].

Therefore, the question arises, what was it about Turgenev that made him so influential in/for Austria during the 19th century? There are some obvious answers to this question. In the wake of Königgrätz's defeat on July 3, 1866, the German Confederation was disbanded, the settlement with Hungary was ratified, and the international position of the Danubian monarchy was weakened. In the aftermath of the defeat, a search for new directions and patterns was carried out on a geopolitical and cultural levels, which, among other things, could have rendered Cisleithania's numerous Slavic crown lands more meaningful. According to Marie von Ebner-Eschenbach, a Turgenevian, the "Austrian muse" at that time was also seeking autonomy from literary movements of imperial Germany. Turgenev's prose was apparently integrated into Austria's target culture, both as an orientation matrix and as a fertile ferment for its production [5].

It is noteworthy that two volumes of Turgenev's stories in German translation by Friedrich Bodenstedt appeared immediately before Königgrätz, in 1864–65, which are thoroughly reviewed by renowned literary figures Hieronymus Lorm [35] and Ferdinand Kürnberger [36] in prominent periodicals in Austria. In his essay, published a year later, Kürnberger sets a very specific challenge for his fellow compatriots, not limiting himself to just praising Turgenev as the Russian Shakespeare of novella. In his opinion, in that period it was essential for Austria to cultivate its own author, with the same awareness of Central and Eastern European nature and the "Slavic soul", which, however, could be expressed through the German language:

> Turgenev, what is he to us? He is Russian. It is not our literature that he belongs to, but only our translation literature. Would it not be better if instead of the Great Russian Turgenev we had a Little Russian, an East Galician, which is an Austrian, i.e. a German? The question is, how would it be, if in this Austria, whose Germanizing mission has so far been so badly fulfilled, if in this time, when the nationalities of Austria are in revolt against Germanism, a Slavic-born poet from the Pruth had to send an excellent German novella to the banks of the Main and the Neckar? [37: 191–192]

It was Leopold von Sacher-Masoch, trained personally by Kürnberger, who became the first "new creation" of this kind, as an "Austrian" or Galician Turgenev. Among those who followed him were Marie von Ebner-Eschenbach, Ferdinand von Saar, and Karl Emil Franzos, who adapted and improved Turgenev's basic pattern of content and form in their novellas and contemporary novels with the binary opposition between village and city like in "Fathers and Sons", as well as, the types of a strong woman ready to sacrifice, a superfluous man, a nihilist, and a hunter-narrator as in "The Hunting Sketches". Despite the fact that we have only mentioned four major names, there were many more followers and admirers of Turgenev in Austria, which will be discussed in more detail.

4.2 Turgenev and the Austrians

In his lifetime, Turgenev visited Vienna only twice, both times briefly and in poor health. The writer's first stay from 7 to 12, April in 1858, on his way from Italy to London, was motivated by the need to consult two renowned Viennese doctors: a well-known physician Johann von Oppolzer (1808–71) and Carl Ludwig Sigmund Ritter von Ilanor (1810–83), the great balneotherapist of that time. During a consultation, the latter advised Turgenev to undergo several weeks of treatment in Karlovy Vary, which he did but only fifteen years later. There was little contact with the Austrian literary scene during that visit. The subsequent trip to London via Paris was even more enriched with new beneficial contacts. During that time, Turgenev met Moritz Hartmann, an Austrian writer, publicist, and translator (see 6.3), who lived as a political refugee in Paris and later, since 1848, in Stuttgart. Hartmann was a tutor in the noble Russian family of Trubetskoy (the wedding of Hartmann's disciple Ekaterina Trubetskaya, who married Turgenev's friend Count Nikolay Orlov in London, was the real reason for Turgenev's trip) [19: 310–316].

The second trip to Vienna was made during the Great World Exhibition from June 12 to 19, 1873. Turgenev had to spend the entire week in his hotel room due to a sudden attack of gout, so he missed exhibition events and the opportunity to establish further contacts. However, he managed to meet briefly with two Austrian acquaintances, Berta Hartmann, the widow of Moritz Hartmann, and Josephine von Wertheimstein (1820–94), a prominent benefactor from Viennese financial bourgeoisie society, also an Austrian acquaintance from Paris [18: 457]. During his subsequent stay in Karlovy Vary, which lasted several weeks, Turgenev was much more productive.

From June 20 to July 24, 1873, Turgenev was in Karlovy Vary for over four weeks, and subsequently in 1874 and 1875 for periods of three and seven weeks. During that visit, he established most of his Austrian contacts.

Considering all of these factors, along with the fact that, between 1863 and 1870, Turgenev maintained the most important and valuable contacts in Baden-Baden, his staying in a spa town raises the question of whether it is relevant to the writers' networks. A visit to a spa town can result in a noticeable increase in contacts for a literary figure [38, 39], as Turgenev's trips to Karlovy Vary demonstrated [40], though this has not yet been a subject for network research.

Since Karlovy Vary was primarily a resort for Vienna's high society in the second half of the 19th century, we focus on Turgenev's Austrian acquaintances. One of the main attractions in Karlovy Vary was the salon of Hermine Seelen (1836–1912), the educated wife of Josef Seelen (1822–1904), local famous balneologist. Hermine also hosted a literary salon in Vienna during the winter and was acquainted with most of celebrities there.

According to Turgenev's letters from Karlovy Vary [18, 19], the writer frequently visited Hermione's salon, where he met the influential director of the Vienna Burgtheater, Heinrich Laube (1806–84), and Betty Paoli (Barbara Elisabeth Glück, 1814–94), the most famous poetess in Austria. In 1874 in Karlovy Vary Turgenev met again with Moritz Hartmann's widow Berta.

5 Turgenev's Austrian Networks

5.1 Turgenev's Indifference to Contacts with Literary Colleagues

Turgenev's Austrian contacts must be examined in their direction and reciprocity to reconstruct the whole network. How did the "great Russian master" relate to representatives of Austrian cultural fields and how were literary contacts handled? First of all, there is a rather disappointing answer to this question. On the basis of Turgenev's academic edition of his letters [18, 19], we have reconstructed Turgenev's Austrian contacts as follows (see Fig. 1).

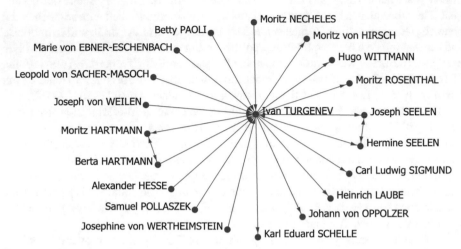

Fig. 1. Ivan Turgenev's Austrian Network. Blue links are non-literary contacts; red links are literary contacts.

There are nineteen contacts in total, but only six are literary (marked in red in the visualization), and only one of them – with Moritz Hartmann – shows reciprocity. The remainder of Turgenev's Austrian connections is limited to contacts with doctors and cultural figures, such as cultural philanthropists Baroness Wertheimstein and Baron von Hirsch. The latter was expected to act as mediator for Turgenev's partner M-me Viardot, or for her son Paul Viardot, beginning his singing career, or for the child prodigy, Moritz Rosenthal, Franz Liszt's student. Moreover, Turgenev's only letter to Laube, a dominant figure in Viennese theater and literature, has a more businesslike, mediatory character, since Turgenev requests a favorable review in the Viennese press on the work by Gustave Flaubert, Turgenev's French friend [19: 418].

Moritz Hartmann. Moritz Hartmann, one of Turgenev's strongest literary ties, is acknowledged not just as an interesting poet and prose writer, which he was, but also as a congenial translator of his stories, especially of the Turgenev's novel "The Smoke". It is notable that Turgenev hardly saw Hartmann as a fellow writer. The latter interested him almost exclusively as a mediator, mainly via translations or critique of his own texts, all the more so because after the amnesty in 1868 Hartmann returned from his exile to Vienna, where for the last three and a half years of his life he headed the powerful paper "Neue Freie Presse". Turgenev's exchange with Hartmann, however, was reciprocal, for the Russian author introduced Hartmann to publishers as a translator and editor under very good financial conditions.

Hermine Seelen. Turgenev was also provided with similar editorial services by Hermine Seelen, who regarded him as a "first-rate writer" [18: 312]. In contrast, Hermine's attempt to communicate with Turgenev about Ferdinand von Saar (one of the most important Austrian Turgenevians) was met with little enthusiasm. He also had some harsh comments about this "twisted" woman in a letter to M-me Viardot [18: 309]. In any case, Turgenev treats Hermine more mildly in the next letter to Viardot, likely because he saw in her a good proofreader for his texts. A letter dated 26.08.1874 informs us that Turgenev engaged her to proofread his "The Hunting Sketches" for the German-language edition of his collected works.

Betty Paoli. Turgenev had no literary dialogue even with the Queen of Austrian Poetry Betty Paoli who tries in vain to contact him in 1869 and sends him a sonnet dedicated to him:

> As a mark of honour, let me send you these greetings,
>
> You're a great master, to whom powers are given.
>
> With sure in hands and no assistance.
>
> To untie this tangled knot of human existence.
>
> (from the collection "Neueste Gedichte", Vienna, 1870).

Probably in Karlovy Vary, Paoli sought to build up more active contact with Turgenev, the fellow writer she admired. In one of his letters to M-me Viardot, his verdict on her sounds harsh: "Mrs. Paoli is exactly as you said: un-bear-ab-le!" [18: 160].

Turgenev was approached in correspondence by representatives of the Austrian literary scene with enclosed works (Marie von Ebner-Eschenbach, Josef von Weilen) or with invitations to cooperate (Leopold von Sacher-Masoch). Despite these attempts, the "meister" responds with polite but brief letters of thanks that do not suggest continued communication. The correspondence between Turgenev and his Austrian admirers was not reciprocated and all of the material objects such as books and manuscripts enclosed in letters were ignored, except for one autograph, which was sent to Alexander Hesse, an autograph collector from Graz.

There is sufficient evidence to demonstrate that Turgenev embodies Pierre Bourdieu's second type of author, a kind of calculus networker: the one who was less interested in the creative-professional expertise of others, even fellow authors ("the pole of pure production, in which producers usually have only other producers as clients (who are rivals as well)" [41: 121]), but rather in the distribution and reception of his own works among the wider public ("the pole of large-scale production, subordinated to the expectations of a wide audience" [41: 121]).

In his role as a notorious "calculus networker", Turgenev consistently ignored the Austrian writers, who were enthusiastic but had little to offer for the distribution of his works to the wider public: no contact was maintained, no cooperation proposal was accepted, and no area of potential common interest was explored.

5.2   Austrian Turgenevians

This, however, does not weaken Turgenev's Austrian fan club. In that case, it was probably due to well-functioning structures within which Turgenev and the Turgenevian works were widely spread, exchanged, and communicated actively in Austria. These structures were mostly reconstructed using the diaries of Marie von Ebner-Eschenbach, which were kept continuously from 1856 until her death in 1916, and some other sources listed in Sect. 3.

The result was the formation of a dense network (see Fig. 2), in which Turgenev's works were read and, more importantly, communicated – recited, discussed, sometimes even criticized. Ties represent the channels through which Turgenev's works were transmitted and communicated. In this case, red marks ties that are particularly important from a literary point of view.

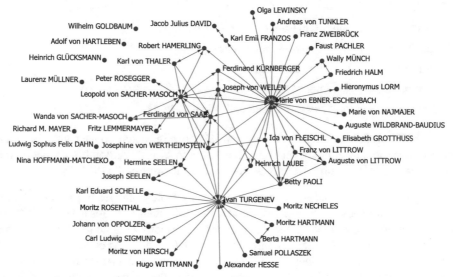

Fig. 2. Network of Austrian Turgenevians. Blue links are non-literary contacts; red links are literary contacts.

The established network proves that Turgenev was actively perceived and communicated within the Austrian cultural field not only in the literary scene, but also in other field segments: in the salons of the old nobles (Ebner-Eschenbach), in theaters (Wally Münch), in publishing houses and newspapers (Kürnberger, Norm, Necheles), in feminist circles (Marie von Majmajer), as well as among higher military officials (Franz von Littrow) and among the rising financial bourgeoisie (Wertheimstein), and Catholic priests (Laurenz Müllner).

The visualization shows, that, firstly, Turgenev's work spread not chaotically and uncontrollably, but in a direct, channeled way. Secondly, in the above-mentioned Turgenevian network, we can see three strong clusters, each centered on Marie von Ebner-Eschenbach, Ferdinand von Saar and Leopold von Sacher-Masoch, all interconnected.

In a complex system like this one, which is hard to cover completely in this article, there are multiple cliques that can be identified. We have chosen two triangles for more detailed consideration: Leopold von Sacher-Masoch – Wanda von Sacher-Masoch – Ferdinand von Saar and Marie von Ebner-Eschenbach – Betty Paoli – Ida Fleischl von Marxow. Of all groups, these two cliques are of particular interest to us, first, because they involve relationships that are both literarily relevant and literarily productive. Secondly, the leading figures of both groups – Leopold von Sacher-Masoch, Betty Paoli and Marie von Ebner-Eschenbach – were also in epistolary (Paoli – also in personal) contact with Turgenev, admittedly, as in other similar cases, without any particular interest on the part of the Russian "master" (see Sect. 5.1).

The Clique Marie von Ebner-Eschenbach–Ida Fleischl von Marxow–Betty Paoli. Based on a visualisation of a network segment, the first clique represents three actors, and it seems quite likely that the above-mentioned poem by Betty Paoli faithfully reflected the atmosphere of Turgenev cult that prevailed in it.

The Clique-Triangel, or "Writing circle" in the relevant literature, of Marie von Ebner-Eschenbach, Ida Fleischl von Marxow (resp. Ida von Fleischl) and Betty Paoli formed in the mid-1860s, just at the time when Turgenev's fame was rapidly rising. At that moment out of three circle participants, only one – Betty Paoli – was already an established great in literature, called "the first lyricist of Austria" by Franz Grillparzer. Ebner-Eschenbach, who today enjoys "status as a national icon," [42: 314] was more of a literary beginner at the time, but quickly developed, not least thanks to constant exchange with both other Circle-members.

The end of this solid community of three literarily important and influential representatives of the Viennese cultural field was the passing of Betty Paoli in 1894 and of Ida Fleischl von Marxow in 1899. For almost 30 years, the regular meeting of three literarily committed friends functioned as a productive writing laboratory, in the context of which they discussed and revised each other's works at regular weekly meetings. Thanks to literary salon evenings in the prestigious well-off house of Ida Fleischl von Marxow, whose husband was president of the Viennese Tramway Society, the clique effectively functioned as an active mediator, being "the core of a much wider network of connections between writers, many of whom were women" [42: 327].

At the time the Writing Circle existed, Turgenev was a literary great par excellence. Consequently, his work and his person occupy a central place in the activities of the clique.

Turgenev's works acted "inside" the clique as important imitation models and stimulators of the own literary creativity of Paoli and Ebner-Eschenbach [20, 21]. "Outward" they were conveyed intensively through Fleischl's salon evenings, where Turgenev's works were often discussed and recited [20].

Since the Russian master gave a cold shoulder to the Circle participants despite their (i.e. especially Paoli's – see Sect. 5.1 and Ebner-Eschenbach's [20]) contact propositions, the whole intensive mediation mechanism of Turgenev's works functioned without his personal participation and without his intervention. The form of communication that three Austrian Turgenevians sought and maintained among themselves – "mutual encouragement and intellectual exchange" [42: 309] – clearly belonged to the first type of writer's exchange according to Bourdieu (see Sect. 5.1), a kind of "romantic poets friendship" [42: 310], while Turgenev, as stated above, cultivated a second, calculating type of communication in the spirit of the realist epoch.

The Clique Leopold von Sacher-Masoch–Wanda von Sacher-Masoch–Ferdinand von Saar. The clique of Turgenevians just considered operated their Turgenev cult consciously and continuously in the course of almost 30 years. The evidence of active communication of Turgenev's works was obvious in the case of three Viennese ladies: in particular Ebner-Eschenbach's diaries bear witness to this [20]. Logically, but only secondarily, we find the traces of Turgenev's influence also in the works of Paoli and Ebner-Eschenbach.

The formation Leopold von Sacher-Masoch – Wanda von Sacher-Masoch – Ferdinand von Saar can be reconstructed as a clique of Turgenevians primarily on the basis of their works. The biographical relations of the actors become obvious in their Turgenev-inspired quality only after a source research. It was established relatively early on that the literary works of all three authors display one and the same plot structure: a weak man is attracted to a woman of strong character, who torments him (morally, or also physically) and in the end takes another, virile and beautiful lover (in "Venus in Furs" this third person is called "Greek"), which increases both the torments of the first protagonist and – in a perverted way – his pleasures and ecstasies. The phantasm and performance nowadays called masochism traces its name back to Leopold von Sacher-Masoch – the very first "Austrian Turgenev" (see 4.1). Also, the masochistic triangular constellation itself, as has been proved, goes back to Turgenev's stories ("An Exchange of Letters", "Spring Waters", etc.). "Masochism should have been called Turgenevism" [43]. However, this paradigm was radicalized under Sacher-Masoch's pen – especially in "Venus in Furs" (1869) and in "Mother of God" (1881).

Wanda Sacher-Masoch, under the influence of and at the coaxing of her husband, also wrote some "masochistic-turgenevist" stories, which were published in the anthologies "Echter Hermelin. Geschichten aus der vornehmen Welt" (1879) and "Die Damen im Pelz" (1883).

We find a milder variant of the same masochostic constellation in the other Turgenevian – Saar. His "strong women" (cf. "Vae victis" (1881), "Ginevra" (1892)) torment the weak male protagonists rather morally, without treating them with a whip, which was the case in the work of both Masochs.

Since Saar was connected throughout his life with rich patronage from the circles of high nobility (Princess Maria zu Hohenlohe [23]) and financial bourgeoisie (the families of Gomperz and von Wertheimstein) and Sacher-Masoch with journalistic circles (Ferdinand Kürnberger, Karl von Thaler and others), their "Turgenevian" lines of communication ran independently of each other rather along these channels (see the visualization) in the form of correspondence or magazine reviews. Strongly Turgenevian in their works, they hardly met in real life on the "Turgenevian" platform, despite the rather fleeting meeting of both men in Vienna in 1873 at the time of the World's Fair (the gout-stricken Turgenev was also in Vienna at that time – see 5.1 – but hardly left the hotel).

A "true-to-life" masochistic, i.e. hereby also "Turgenevistic" constellation occurs with both Sacher-Masochs and Saar in three months (August–October) of 1876, when they met in the bathing resort Bad Frohnleiten. Sacher-Masoch, who was inclined to masochistic performance in life as well, had Wanda staged in the ambience of the spa as a devised dominatrix and man-eater: "playing billiards in a decollete jacket in the spa house, smoking cigarettes, being courted by young men" [44]. His goal was to find a suitable "Greek" for a triangular masochistic game, which did not work out. Saar alone, as a former Guard officer and "handsome, splendid man," as Wanda characterizes him in his memoirs, seemed to fit. "He has the looks to match the Greek," said Leopold, "alone he is far too much of a poet for the role." [44]. Whether a "real" masochistic enactment occurred during the said three months or the exchange of three Turgenevians remained purely communicative has remained unknown. What is certain in any case is that the stay in Bad Frohnleiten was beneficial to Sacher-Masoch's productivity as an author: "During this time, my husband wrote the most beautiful and the best" [44], according to Wanda in her memoirs.

Twenty years later, Saar represented Leopold von Sacher-Masoch in his novella "Ninon" (1897), as a prematurely aged writer Z., and his wife Wanda, as the men-eater Ninon. It is also noteworthy, that Turgenev is referred to in the text as "a foreign writer":

Yes, his bright, colourful creations could not be compared with anything that had gone before. It is true that one wanted to recognize the influence of a foreign writer in them. […] It was always the same love story: the same weak, will-less man writhing in the dust – and the same ruthless, cruel, brutal woman. [45: 300–301]

The masochistic-turgenevistic constellation – a strong cruel woman and a weak humiliated man – is also mentioned here, admittedly without the "third", a "Greek", whose role the Sacher-Masoch couple is said to have once attributed to Saar himself.

If one takes a look at both cliques in the context of the whole Turgenevian network, it becomes obvious that they are connected to a larger formation mainly via the tie Ebner-Eschenbach – Saar, who communicated intensively with each other on a personal and epistolary [22] level throughout their lives. The SNA approach helps to see, apart from such superordinate structures, even smaller, subordinate ones. When entire chains of communication are reconstructed, otherwise hidden nodes come into view. This is the case, for example, with Ida Fleischl von Marxow, whose status as a Turgenevian was hardly considered earlier.

6   Conclusion

There are three conclusions to be drawn from this study.

1. A network approach and visualization of Turgenev's personal contacts help reveal, concretize and clarify the inner structure of Turgenev's contacts, while re-metaphorising his image as a "master of ties".
2. Studying the network of Austrian Turgenevians in our context may also contribute to reviewing the general pattern by using the method of network analysis. Historically, Turgenev's literary international relations have been analysed on the basis of the names of rank; however, studying the network of Austrian Turgenevians gives us a different perspective. According to Lore Knapp: "When … a network of actors is traced, new perspectives on historical connections emerge that allow to inter-pret existing value systems and hierarchies in the field of the history of literature

and aesthetics. The tracing of connections makes nodes to emerge that are otherwise obscured by the established canon-specific hierarchies or a historiographical high-altitude aesthetic" [15: 20]. The productivity of "the tracing of connections" is confirmed in particular by the second visualization, which reveals a long (probably far from complete) series of Austrian Turgenevians whose names would hardly be reconstructible in this context without SNA procedures.

3. As demonstrated by the "self-organized" network of Austrian Turgenevians, which was formed "despite" the author's disinterest, a network as a "cultural technique" [46: 8] follows its own laws. SNA helps illustrate this inner pattern like no other method, and thus becomes an indispensable research tool to complement cultural transfer and comparative literature analyses.

Acknowledgment. This study was supported by a grant from the Russian Science Foundation (RSF) No. 22–28-01186 entitled "The resort as a topos, network, and narrative in European literature of the modern era".

References

1. Thiergen, P.: Ivan Turgenev: "Dvorjanskoe gnezdo (Das Adelsnest)". In: Zelinsky, B., (ed.): Der russische Roman, Böhlau, Wien, 164–185 (2007)
2. Gerigk, H.-J.: Turgenjew. Eine Einführung für den Leser von heute, Winter, Heidelberg (2015)
3. Gerigk, H.-J. (ed.): Turgenjew – der russische Europäer: Fünf Vorträge der Turgenjew-Konferenz 2016 in Baden-Baden. Mattes Verlag, Heidelberg (2017)
4. Cheauré, E., Nohejl, R., Gorfinkel, O., (ed.): Russland in Europa – Europa in Russland. 200 Jahre Ivan Turgenev. Begleitbuch zur Ausstellung im Stadtmuseum Baden-Baden 22. September 2018 bis 3. März 2019. Stadtbibliothek, Baden-Baden (2018)
5. Poluboyarinova, L.: "A teper' ešČo i Turgenev!". Istoki, osnovanija i ključevye parametry recepcii russkogo klassika v Avstrii ["And now Turgenev!" Origins, foundations and key parameters of the reception of the Russian classic in Austria], SPBU Publishing, St. Petersburg (2018)
6. Figes, O.: The Europeans: Three Lives and the Making of a Cosmopolitan Culture. Metropolitan Books Henry Holt and Company, New York (2019)
7. Polubojarinova, L., Frick, W., von Essen, G., Hauser, K., Kulishkina, O.: Ivan Turgenev als Netzwerker: Digitale Kuratierung seiner europäischen, insbesondere deutschen literarischen Kontakte (am Beispiel seines Briefwechsels vom Juni 1868 bis Mai 1869). Seminar: A Journal of Germanic Studies. 57, 3, 217–242 (2021) https://doi.org/10.3138/seminar.57.3.2
8. Poluboyarinova, L., Kulishkina, O.: Frames and networks: framework narratives in 19th-century European literature. In: Antonyuk, A., Basov, N. (eds.) NetGloW 2020. LNNS, vol. 181, pp. 49–65. Springer, Cham (2021). https://doi.org/10.1007/978-3-030-64877-0_4
9. Espagne, M.: Quelques aspects actuels de la recherche sur les transferts culturels. In: Arnoux-Farnoux, L., Hermetet, A.-R., (Ed.): Questions de réception. Nimes, Champ social Editions, pp. 163–190 (2009)
10. Keller, T.: Kulturtransferforschung: Grenzgänge zwischen den Kulturen. In: Moebius, S., Quardflieg, D., (Ed.) Kultur. Theorien der Gegenwart, 2nd ed., Wiesbaden, Verlag für Sozialwissenschaften, pp. 106–117 (2006)

11. Hoff, K., Schöning, U., Weinmann, F. (ed.): Internationale Netzwerke: literarische und ästhetische Transfers im Dreieck Deutschland, Frankreich und Skandinavien zwischen 1870 und 1945. Königshausen&Neumann, Würzburg (2016)
12. Scott, J.: Social Network Analysis, 2nd edn. SAGE Publications, London, Thousand Oaks, New Delhi (2000)
13. Scott, J., Carrington, P.J., (ed.): The SAGE Handbook of Social Network Analysis. SAGE, London, Thousand Oaks, New Delhi (2011)
14. Edmondson, Ch., Edelstein, D. (eds.): Networks of Enlightenment: Digital Approaches to the Republic of Letters. Oxford University Studies in the Enlightenment, Liverpool University Press (2019)
15. Knapp, L. (ed.): Literarische Netzwerke im 18. Jahrhundert, Aisthesis Verlag, Bielefeld (2019)
16. Thomalla, E., Spoerhase C., (ed.): Werke in Relationen. Netzwerktheoretische Ansätze in der Literaturwissenschaft (= Zeitschrift für Germanistik. Neue Folge, 29,1) (2019)
17. Rollinger, C.: Prolegomena: problems and perspectives of historical network research and ancient history. J. Hist. Netw. Res. **4**, 1–35 (2020) https://doi.org/10.25517/jhnr.v4i0.7225
18. Turgenev, I.: Polnoe sobranie sočinenij i pisem [Complete collection of works and letters]. In: 30 Vol. Pis´ma [Letters]. In 18 Vol., Vol. 12. 2nd ed. Nauka, Moskau (1982–2020)
19. Turgenev, I.: Polnoe sobranie sočinenij i pisem [Complete collection of works and letters]. In: 30 Vol. Pis´ma [Letters]. In 18 Vol. ,Vol. 13. 2nd ed. Nauka, Moskau (1982–2020)
20. Golz, J.: Tagebücher. In: Witte, B., Schmidt, P. (eds.) Goethe Handbuch, pp. 396–409. J.B. Metzler, Stuttgart (1997). https://doi.org/10.1007/978-3-476-03654-4_15
21. von Ebner-Eschenbach, M., von Knorr, J.: Briefwechsel 1851–1908. Akademie-Verlag, Berlin (2016)
22. von Saar, F., von Ebner-Eschenbach, M.: Briefwechsel. Wiener bibliophilen Gesellschaft, Wien (1957)
23. Hohenlohe, M.Z., von Saar, F.: Ein Briefwechsel, Druck und Verlag von Christoph Reisser's Söhne (1910)
24. Österreichisches Biographisches Lexikon. http://www.biographien.ac.at/oebl?frames=yes
25. Goldbaum, W.: Turgenjew's deutsche Jünger. Eine kritische Randglosse. Mehr Licht. Eine deutsche Wochenschrift für Literatur und Kunst **27**(1), 424–425 (1879)
26. von Thaler, K.: Nihilismus in Deutschland. In: Farin M. (ed.): Leopold von Sacher-Masoch. Materialien zu Leben und Werk, Bouvier, Bonn, pp. 43–51 (1987)
27. Glagau, O.: Turgeniev's Nachahmer. Karl Detlef – Sacher-Masoch. In: Glagau, O.: Die russische Literatur und Iwan Turgeniev, Gebrüder Paetel, Berlin, pp. 162–174 (1872)
28. Storm, T.: Briefe: In 2 Vol, vol. 1. Aufbau-Verlag, Berlin-Weimar (1972)
29. Wytrzens, G.: Zur österreichischen Turgenjew-Rezeption bis 1918. Wiener Slawistisches Jahrbuch **10**, 107–126 (1982)
30. Hanisch, O.: Die Turgenev-Rezeption im Werk Ferdinand von Saars. Diss. Magdeburg (1987)
31. Polubojarinova, L.: I.S.Turgenjews "Väter und Söhne" in österreichischer Rezeption. In: Arlt H. (ed.): Kunst und internationale Verständigung, Röhrig, St. Ingbert, pp. 140–151 (1995)
32. Hellberger, M.: Tierdarstellungen als symptom des gesellschaftlichen strukturwandels im interkulturellen kontext: die rezeption von Ivan Turgenevs Tierdarstellungen im Werk Marie von Ebner-Eschenbachs und Ferdinand von Saars. Holzner J. (ed.): Russland – Österreich. Literarische und kulturelle Wechselwirkungen, Peter Lang, Bern, Berlin, 97–126 (2000)
33. Polt-Heinzl, E.: Was aber ist modern? Arthur Schnitzler und Peter Altenberg vs. Ferdinand Saar, Elise Richter oder Rosa Mayreder. In: Taschkenov, S., (ed.): Visionen der Zukunft um 1900: Deutschland, Österreich, Rußland, Fink, München, pp. 135–158 (2014)
34. Tanzer, U.: Konzeptionen des Glück im Werk von Marie von Ebner-Eschenbach. Semin.: J. Germanic Stud. **47**(2), 254–267 (2011). https://doi.org/10.3138/seminar.47.2.254
35. Lorm, H.: Iwan Turgenjews Erzählungen. Österreichische Wochenschrift für Wissenschaft, Kunst und öffentliches Leben. Beilage zur kaiserlichen Wiener Zeitung, 4, 1643–1647 (1864)

36. Kürnberger, F.: Turgénjew und die slawische Welt. In: Kürnberger, F.: Literarische Herzenssachen: Reflexionen und Kritiken, Karl Proschaske, Wien, Teschen, 106–121 (1877)
37. Kürnberger, F.: Vorrede zum „Don Juan von Kolomea". In: Sacher-Masoch, L. von: Don Juan v. Kolomea. Galizische Geschichten, Gruyter, Bonn, pp. 189–192 (1985)
38. Borowka-Clausberg, B.: An den Quellen des Hochgefühls: Kurorte in der Weltliteratur. In: Eidloth, V. (ed.): Europäische Kurstädte und Modebäder des 19. Jahrhunderts/European health resorts and fashionable spas of the 19th century, Univ.-Bibl., Heidelberg, Stuttgart: Theiss, pp. 217–230 (2012) URL: https://journals.ub.uni-heidelberg.de/index.php/icomoshefte/art icle/download/20355/14149
39. Soroka, M.: Spa society in Russian literature, 1829–1912. Forum for Modern Language Studies. 2019. Vol. 55, No. 4, p. 426–443. https://doi.org/10.1093/fmls/cqz034
40. Nazarova, L.N.: Turgenev v Karlsbade [Turgenev in Karlovy Vary]. Turgenevskij sbornik 2, 279–286 (1966)
41. Bourdieu, P.: The rules of art: genesis and structure of the literary field. Stanford University Press, Redwood (1995)
42. Watzke, P.: Friedenship and Networking: The Schreibzirkel von Marie von Ebner-Eschenbach, Ida von Fleischl-Marxow and Betty Paoli. In: Byrd, V., Malalakaj, E. (ed.): Market Strategies and German Literature in the Long Nineteenth Century. De Gruyter, Berlin, Munich, Boston, pp. 307–332 (2020)
43. Finke, M.C.: Sacher-Masoch, Turgenev, and Other Russians. In: Finke, M.C., Niekerk, C., (ed.): One Hundred Years of Masochism. Literary Texts, Social and Cultural Contexts. Rodopi, Amsterdam, Atlanta, pp. 119–137 (2000). https://doi.org/10.1515/9783110660142-013
44. von Sacher-Masoch, W.: Meine Lebensbeichte. Memoiren. https://www.projekt-gutenberg. org/sacherwa/memoiren/chap004.html
45. von Saar, F.: Ninon. In: von Saar, F.: Werke. In: 3 Vol., Vol 2. Amandus Wien, pp. 299–348 (1959)
46. Gießmann, S.: Die Verbundenheit der Dinge. Kulturverlag Kadmos, Berlin, Eine Kulturgeschichte der Netze und Netzwerke (2016)

Online Networks

Defining the Real Structure of the City Through Spaces of Everyday Activity Based on User-Generated Online Data

Oleksandra Nenko[1]([⊠]), Marina Kurilova[2], Artem Konyukhov[2], and Yuri Bogomolov[3]

[1] Turku Institute for Advanced Studies, University of Turku, Turku, Finland
oleksandra.nenko@utu.fi
[2] Calgary, Canada
[3] Masaryk University, Brno, Czech Republic

Abstract. This paper addresses the question of defining the "real" structure of a city, which accommodates existing activity patterns and facilitates urban management and distribution of the city resources, since this structure does not always correspond to the city administrative division. The motivation for the paper is to present a method of defining spaces of everyday activities (SEAs). Here we define the structure of the city based on user-generated online data, which provides evidence about the usage of the city space in the course of everyday consumer practices. Based on Google Places data for St. Petersburg, we build user-venue and venue-venue networks, where users and venues are connected through check-ins and venues are connected through common users. We develop a clusterization technique to define the units of the geographically proximate and socially similar venues – SEAs. We compare the map of such units with the administrative division of St. Petersburg on two different scales: at the level of agglomeration and at the level of administrative districts. We describe the formation of borders and connections that cause matches and inconsistencies between administrative boundaries and SEAs, such as natural and artificial barriers, transportation and infrastructural connectors.

Keywords: Space of everyday activity · Real structure of the city · User-generated geolocated data · Urban networks

1 Introduction

Motivation for this paper is defining the "real" structure of the city as people-generated, different from the one officially imposed by the decision makers, and as dynamic, reliant on urban life processes, different from the static one fixed in urban planning documents. Knowledge about the real structure of the city gives an opportunity to plan the relational (rather than top-down) distribution of the city's limited resources reliant on the clusterization of citizens' needs in the city space. It enhances the practices of fighting the hidden segregation of the city space not evident in the official aggregated economic and

M. Kurilova and A. Konyukhov—Independent Researcher.

© The Author(s), under exclusive license to Springer Nature Switzerland AG 2023
A. Antonyuk and N. Basov (Eds.): NetGloW 2022, LNNS 663, pp. 97–113, 2023.
https://doi.org/10.1007/978-3-031-29408-2_7

social indexes [3, 8, 21] and rooted in the inequality of opportunities on the personal level. It introduces the "human" perspective into the city management, dependent on the acknowledgement of the local and changing demands. Decentralisation of urban management and relocation of decision-making functions to local bodies of self-government are also important ways of democratising and humanising the development of the urban space [32].

Understanding the actual borders of "living units", urban management bodies can refine the system of allocating local community facilities, which are usually distributed according to administrative division. These community facilities include schools, hospitals, kindergartens, public spaces, and transportation points, the number of which is defined according to the population size of an administrative unit. However, this approach to calculation does not take into account human patterns of spatial behaviour, which might be influenced by natural and physical barriers inside the unit or the attractors beyond it. Hence the borders of the administrative unit do not delineate the real connections of people to the facilities. This can lead to insufficient service provision in areas with scarce services and additional load on the service infrastructure in areas with overconsumption. In this paper we aim to show an approach to define the real units of human activity, delimited by usage of community facilities.

We relate to the term "local activity space" introduced by Schnell and Yoav in 2001 and defined as a territory of everyday life usage that reflects the recurrent patterns of human behaviour in-between the working and home place [26]. We are interpreting this term as a "space of everyday activity" (SEA) – geographical areas defined by patterns of everyday consumption. The relational character of this process is manifested in connections formed between users and the facilities, such as cafes, shops, and parks, when the former visit the latter. Given this relational nature of geography of consumption, spaces of everyday activity can be analysed through user-generated and geolocated online data coming from location-based social networks, as we have shown earlier [15].

Recent advances in the sphere of applying user-generated and geolocated data to spatial analysis are informing a new turn in the analysis of the real structure of the city [7]. We follow this trend in introducing a new method to process user-generated data and demonstrate it using the case of St. Petersburg, Russia.

In the result we introduce an approach to analyse the real structure of the city space adapted for the aim of better management of city resources based on data generated by users online in location-based social networks.

2   Literature Review

Classical studies on vernacular areas in cities and regions have showed the differences in naming [22, 23], bordering [2], and classifying different units of the city that have been formed by people of different ages, genders, lifestyles, and social backgrounds living in different urban or regional contexts. Studies on vernacular areas have had a tremendous impact on spatial planning, such as navigation improvement [27] or detecting agglomeration economies [4]. Vernacular areas are also invoked as sites of resistance, local-level organising, and activism in defiance of the dominant policy agenda [12, 18, 24, 25].

The term SEA is different from the traditional geographical term of the "vernacular area", as it is an objective representation of human practice and not a mental representation of an area.

One of the basic points of departure in our paper is that everyday life practices define the pattern of urban space usage and actual tracks and borders of the city unit "lived" by the person. In his seminal work, Lynch has demonstrated the importance of the perceived structure of urban space which depends on its mental representation and is linked with recurrent spatial practices [11]. As an urban planner, Lynch considered a residential area to be a basic planning unit physically and mentally, and believed that the borders of the residential unit must encompass all the necessary elements of the service infrastructure. Good basic units have strong borders recognized by people and reinforced through their practices; units without an understandable system of paths, nodes and landmarks, distorted by borders, lose their cohesion and attraction to people.

Sociological theories adhering to the critical reception of urban space as a relational and changeable phenomenon constructed through everyday practices of people guided by different visions of the city highlight the necessity of tracing the patterns of space usage and criticise the administrative images created by decision makers as divorced from real life [5, 10].

Economic and human geographers argue that an understanding of space, especially of its dynamic features, such as the transition from one economic and social situation to another, must be considered through novel flexible conceptual schemata that correspond to such a fluidity. For example, Aksyonov analysed the post-soviet transitions in the character of space usage and distribution of the urban infrastructure in Russian cities, including St. Petersburg, introducing the concept of "spatial-temporal system" which refers to any kind of urban economic entities, e.g., retailers [1]. He shows the transformation of the spatial-temporal system of retail trade over time, which since the early (illegal) market stage in the 1980s has followed the laws of demand rather than the administrative distribution. The fluidity in users' demand for location-based city services, such as transportation stops and routes, in the course of a week and even a day, is shown in studies of spatiotemporal accessibility and demand for services [16]. The availability of basic city services close to one's homeplace and comfortable conditions of their provision (e.g. working hours) have been found to be significant factors of the quality of life [9, 17, 30, 31].

Big data reflects an endless variety of socio-spatial transactions and offers an opportunity to advance the toolkit of urban studies by providing an understanding of the existing partitioning of the space through communication ties, consumer practices and commutation routes [6, 19, 20, 29]. For this reason, based on the standpoints highlighted by previous researchers, we are presenting a method for defining the socio-spatial structure of the city based on Location-Based Social Network (LBSN) data, which reflects the most popular venues people use and the borders of the spatial units they actively engage with.

3 Methodology

The methodology of the paper is based on socio-spatial network analysis of connections between venues and users who visit them. Geolocated check-in data allows tracing connections between users and venues, where a link is formed by the fact of the check-in. Our methodological approach was presented in detail elsewhere [15]. In this paper we are employing a dataset of check-ins from Google Places St. Petersburg and use a refined algorithm of the socio-spatial clusterization. The methodology includes the following steps:

1. Collecting a sample of georeferences check-ins.
2. Construction of the bimodal network (venues-users).
3. Projecting the network "venues-users" into the network "venues-venues". Edge weight between nodes indicate the number of common users.
4. Defining a backbone graph in the "venues-venues" network.
5. Community detection in the backbone graph.
6. Comparing clusters with the borders of the city and borders of administrative districts.

To build a socio-spatial network, we have collected a database of venues from Google Places[1] and a database of users' check-ins and comments on the same venues from Google Maps. The venues in Google Maps and Places databases cover open-air and indoor public spaces (parks, libraries, cafes, etc.), as well as urban services representing different areas of urban infrastructure (healthcare, sports, education, transportation). In our database there was no limitation for the functional type of venues included. Google Places provides access to information about venues in a city through a paid "REST" API, and for a city the size of St. Petersburg the cost of the data would amount to thousands of US dollars. To eliminate the cost of the dataset, we have used a multistage nested

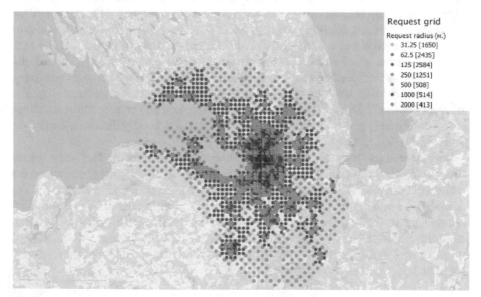

Fig. 1. The boundaries of the sample against the administrative borders of St. Petersburg (the number of sampled venues in a given area is given in square brackets in the legend).

---

[1] Google places web: https://cloud.google.com/maps-platform/places/.

sampling method. The sampling frame was set to include the settlements close to the city borders (Murino, Devyatkino, Krasnoye Selo, Kudrovo, and others), but not officially in them, to detect possible connection of such borderline settlements with the city service infrastructure (Fig. 1).

The geographical area of St. Petersburg was divided into square cells with a side of 2 km in the periphery of the city where the number of venues is smaller, up to cells with a side of 32 m in the city centre, where the number of venues is bigger. The minimal number of venues in the cell is 20 venues. The resulting database consists of 184,759 venues and 2,648,965 check-ins from 933,645 users.

The database was transformed into a bimodal network of "users-venues", where users are connected with venues through the fact of a user leaving a check-in for a venue. For further analysis we have transformed a bimodal network into a one-mode network "venues-venues". The venues in the latter network are connected through the fact of their visitation by the same users. The projection of a bimodal into a one-mode network is done through a transposition of a bimodal matrix and multiplying the bimodal and the transposed matrices [13]. The edges of the network were validated for the statistical significance to avoid projection procedure mistakes through a procedure of defining a backbone graph [14]. The backbone graph was calculated using the "dual-degree conditioned thresholds" method by comparing the original graph with 1,000 random graphs with fixed degrees. The edge between two venues is deemed statistically significant at the specified a-level and counted as present in the backbone if the weight of the original edge is larger than computed in more than $a * N$ random bipartite projections (here $N = 1000$). In this study $a = 0.99$.

To detect the clusters of the venues, the Combo algorithm was applied, which is based on a modularity optimization technique and has proved its quality of detection compared to classical algorithms [28]. Combo unites classical community detection strategies and performs for each source community the best possible redistribution of all source nodes into each destination community and the best merger/split/recombination.

To compare the emerging real structure of the city with the administrative division for various scales of data aggregation, we have conducted clustering procedures on two scales:

– bigger clusters without limitation for geographical proximity of venues were put in a cluster that demarcates the division of the city and the territories at its borders into macro-districts; this scale reveals the real structure of the agglomeration formed around St. Petersburg;
– smaller clusters with a limitation for geographical proximity of venues in a cluster of 2,250 m, comparable with the division of the city into administrative districts ("rayon"); this scale reveals stronger and/or weaker socio-spatial clusters not evident at the first scale.

The first scale of clustering yields 30 SEAs, 14 of which have distinct spatial boundaries (Fig. 2). The remaining 16 are formed by 10% of the venues that are scattered out of the borders of the city and not connected with the rest of the clusters. We propose to use this scale of SEAs to analyse the city in its administrative borders in a wider context; in other words, this is a tool to consider the real structure of an agglomeration.

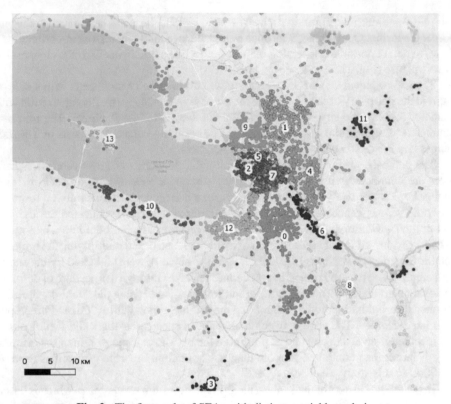

Fig. 2. The first scale of SEAs with distinct spatial boundaries

The second scale of clustering yields 32 smaller clusters, which are geographically more dense and have higher social similarity in terms of users (Fig. 3). To obtain such clusters we have removed the edges from the backbone graph with distances bigger than a threshold number, chosen based on a median distribution of distances between venues, namely 2,250 m. Distances above this value do not affect the results of clustering, while distances below this value lead to the formation of smaller clusters. Clustering results at the first and second scales are hierarchical: all venues in smaller SEAs (32) belong to bigger SEAs (14), with the exception of overlapping SEAs 5 and 7 on the first scale in the city centre, which transform into adjacent SEAs 13 and 14 on the second scale. We argue that the second scale of SEAs can be used to consider localisation of borders and connections between them.

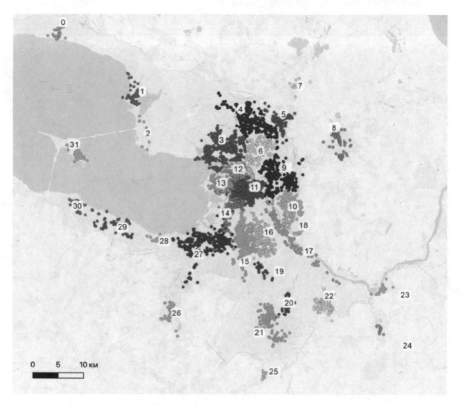

Fig. 3. The second scale of SEAs

4 Results

4.1 The First Scale of SEAs

The resulting first-scale clusters show the real structure of the agglomeration of the city, connectivity, and autonomy between the administrative districts of St. Petersburg and the adjacent towns of the Leningrad region. Figure 4 shows the agglomeration effect of the city in attracting the users of the settlements beyond the city borders (Murino, Kudrovo, Otradnoe, Kolpino) to the venues and services inside the city borders. The map of the first-scale SEAs depicts how these formerly rural settlements, quickly developing into urbanised residential territories, are actually becoming real parts of St. Petersburg through their connection with the city service infrastructure. This illustration provides evidence that these settlements, though increasing in the number of houses, are developing as sleeping quarters and experience shortage of educational, recreational, health, and other services. Consumers are eager to replenish this shortage using the services in the neighbouring city districts, creating clusters of urban consumption overcoming administrative borders. This eventually might lead to unwanted consequences of "abusing" the city services at the administrative margins.

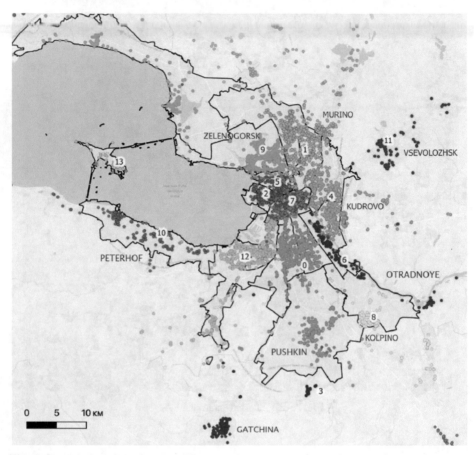

Fig. 4. Boundaries of the first scale SEAs against borders of the administrative districts of the city

Apart from urban planning drawbacks, the first scale of SEA is proving emergent polycentricity in the Leningrad region: for example, Vsevolozhsk and Gatchina, towns beyond the borders of St. Petersburg, establish clusters of their own. However, Pushkin, the historic and full-fledged city on the border of St. Petersburg, has become part of the SEA 0, this showing that its dwellers are extensively using services of the semi-central administrative districts of St. Petersburg (Moskovsky and Frunzensky). Compared to Pushkin, Kolpino (SEA 8) on the city border forms a SEA of its own, connecting together surrounding smaller settlements of Krasny Bor, Ulyanovka, Tosno, and Nikolskoye. Kolpino has lesser transport connectivity with St. Petersburg than Pushkin, what might be the reason for lesser connections.

On the first scale of SEA "strong" borders and strong connections of territories on a city scale become evident. There are cases of classical city sprawl along transport lines, for example several settlements are tightly linked with city districts by the railroad (Pushkin in SEA 0, Krasnoye Selo in SEA 12, Otradnoe in SEA 6). At the same time, remote towns of the Leningrad region, despite their railroad connection, form SEAs of their own, such as Gatchina and Vsevolozhsk.

The connection of city districts into one SEA through the subway system, which has a big impact on the distribution of venues, can also be detected. This is exemplified by Krasnoselsky and Kirovsky districts, which belong together to SEA 12. There is no single subway station in Krasnoselsky district, so the dwellers use the subway station located in Kirovsky district (Prospekt Veteranov station is one of the most used in Russia: it services 83,800 people per day).

The towns of Petrodvortsoviy district of St. Petersburg at the southwestern part of the Gulf of Finland – Lomonosov, Peterhof, Strelna – are forming a unified SEA 10. The towns of Sestroretsk and Zelenogorsk in the north-western part of the Gulf of Finland are not consolidating into a separate cluster, but belong with Primorsky district of St. Petersburg in SEA 9. This difference might be interpreted by the fact that northwest is used as a recreation area for city dwellers, while each of the southwestern towns has become a distinctive tourist destination.

The first scale of SEA shows the connections and alliances between several administrative districts of St. Petersburg, and the borders of all districts can be unambiguously compared with the borders of SEAs. The exceptions are SEAs 5 and 7, which overlap in the historical city centre (Fig. 5). These clusters occupy the Tsentralny, Admiralteysky, and Petrogradsky districts and the Speak of Vasilevsky Island belonging to the Vasilevsky

Fig. 5. The centre of St. Petersburg with SEAs of the first scale

district. However, they have different functional loads and therefore different consumption patterns behind them and user-venue connections dividing them into different SEAs. 80% of the venues in SEA 5 consist of cafes, bars and restaurants, so it can be seen as a cluster of weekend recreation for the city dwellers, while SEA 7 largely comprises cultural institutions, catering and retail services, including souvenir production, resembling a pattern of tourist behaviour (Fig. 6).

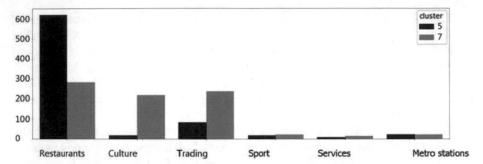

Fig. 6. Ratio of functions of venues in the first-scale SEAs 5 and 7 (based on Google Places categories of venues used in check-ins)

4.2   The Second Scale of SEAs

The second scale of SEAs reveals smaller, more geographically proximate and socially similar clusters (Fig. 7). This scale is comparable to the official division of St. Petersburg into administrative and municipal districts. The spatial peculiarities of human behaviour in a given built environment context become more evident. Several detected patterns of SEAs' formation are described below.

Concurrency of SEA Borders with Administrative District Borders. The venues forming SEAs 15 and 16 neatly fit into the boundaries of the Moskovsky and Frunzensky districts. The contemporary borders of these districts were formed in 1961, separated by railways in the west and east, delineated by vast industrial areas in the north and formed along two highways: Moskovsky prospect and Buharestskaya Street. Both districts are characterised by a well-developed service infrastructure: there are many cultural and leisure institutions, three large parks, and good transport infrastructure, including six subway stations. These are self-efficient territorial units where residents can consume all the needed services within the district.

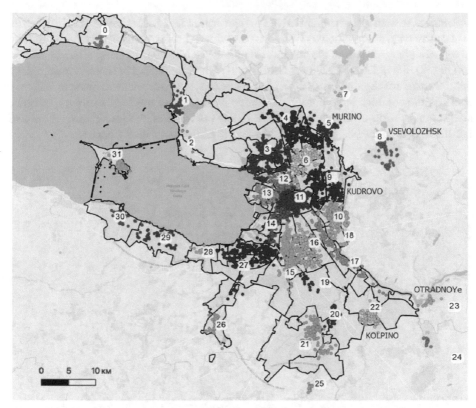

Fig. 7. Boundaries of the second scale SEAs against borders of the municipal districts of the city

SEA 31 is located in the city of Kronstadt on the island of Kotlin (administratively – Kronstadtsky district), which is a completely autonomous entity, created by natural barriers and good internal infrastructure. SEA 22 covers the Kolpino settlement, replicating the cluster on the first scale.

At the second scale of SEAs Petrodvortsovy district is made up of SEAs 30 (Lomonosov), 29 (Peterhof) and 28 (Strelna), which are separate towns, while at the first level altogether they made up SEA 10, coinciding with administrative borders of the district. The Petrodvortsovy district is an "ideal model" in terms of administrative division; it completely coincides with the cluster at the first scale, and second scale clusters completely coincide with the division into municipal districts.

Coincidence with an Administrative District Apart from a Detached Territory. The city centre on the second scale of SEAs demonstrates new patterns of service usage. The two overlapping clusters of the first level separate and form two adjacent SEAs 11, 12 and 13 (Fig. 8). SEAs 12 and 13 almost correspond to their municipal borders – Petrogradsky and Vasileostrovsky districts, respectively – occupying two islands. However, SEA 11 exceeds the administrative borders and breaks off parts of the Vasileostrovsky district at the conjunction of the clusters. SEA 11 covers most of the iconic attractions and facilities, and, in fact, can be named "the heart" of the historic city centre.

Fig. 8. St. Petersburg city centre with SEAs of the second scale

Fragmentation within Administrative District Borders. SEAs' configuration demonstrates how natural and human-made barriers break up one administrative district into several real ones. A vivid example here comprises SEAs 10, 17 and 18 in the Nevsky district – the only district in St. Petersburg located on the two banks of the Neva River (Fig. 9). The division of SEAs of the left and the right river banks is present at both scales. Additionally, an industrial zone creates a significant barrier between Sosnovka (10) and Rusanovka (18).

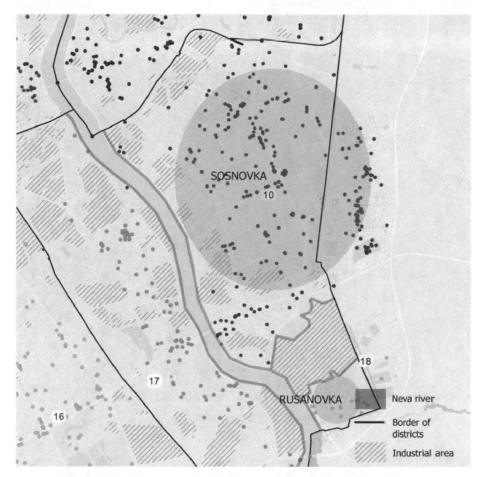

Fig. 9. SEAs 10, 17 and 18 within the boundaries of one municipal district

Exceedance of the Administrative District Borders. Exceedance of several districts by SEAs is provoked by the structure of subway lines in the north and in the southeast of St. Petersburg. For instance, SEA 27 unites two districts – Krasnoselsky and Kirovsky through the subway station Prospekt Veteranov described above. SEA 17 goes from the city centre to the city border along the bank of the Neva River, covering several subway stations of the "Green" line, and thus exceeds the Nevsky district (Fig. 10).

Fig. 10. SEA 17 along the subway line

5 Discussion

The approach presented here provides an illustrative approach to reflect the real structure of the city through consumer behaviour in comparison with the administrative structure of the city. This approach could be used for rethinking the administrative structure of the city to make it more coherent with the real one. However, it should be sensitive to case specifics: it is important to distinguish when the borders of an administrative unit should be redefined and when the borders may be kept, but the interior barriers for human behaviour removed.

Reconsidering administrative borders implies dealing with political and economic issues, such as conflicts of interests between administrative units. For St. Petersburg, such a conflict may take place at the level of city-region relations (St. Petersburg and Leningrad region are independent entities), and at the level of administrative and municipal districts within the city borders.

Inclusion of borderline settlements into the official city borders is also an open question, while it might be also considered that these areas need more self-sufficiency in terms of urban services and might stay in the framework of regional budget rather than city budget. Here the balance between the city's sprawl and formation of a polycentric urban region should be found.

The dataset used in this paper to define the real structure of the city is one of the most voluminous collections available, as it is produced using one of the most popular LBSNs in Russia – Google Maps. However, for a deeper analysis of consumer behaviour in space a more exhaustive account can be done using more relevant sources, one of them being Instagram as a popular network for sharing subjective evaluations and images of places.

6 Conclusion

The analysis of SEAs provides the possibility to uncover the real structure of the city and to compare it with the existing administrative division based on the data about human consumption behaviour. Voluminous user-generated online data from Google Places proves to be spatially representative for the whole city and its agglomeration and insightful for tracking the patterns of service infrastructure usage reflecting clusterization of the city space. The data and approach used illustrate the connections and borders between the clusters of human activity, which are correlated with the items of the built environment. Therefore, the analysis allows us to discover good examples as well as mistakes in urban planning. The presented approach is also sensitive to variation between cases, because it reveals specific clustering of human everyday activities in areas with different urban planning features.

The approach accommodates different scales of analysis, with clear hierarchical structure, complementary to each other. Differentiation between different scales of SEAs offers the possibility to trace spatial trends of different scales – on the scale of agglomeration and on the scale of geographically and socially proximate city territories.

Thus, for St. Petersburg, the first scale of SEA reveals the agglomeration effect of the city and the connection of the borderline settlements with the city in terms of service consumption. On the second scale of SEA several patterns were defined. Full coincidence of clusters with administrative borders (13 SEAs inside the city borders) demonstrates that during planning all natural and artificial barriers were taken into account and sufficient service infrastructure was designed. Mismatch between the boundaries of the SEAs and the administrative division (14 SEAs inside the city borders) shows environmental and artificial barriers that were underestimated or ignored, including water objects, forest zones, industrial areas, speed highways, and railway fines. It also shows the existence of powerful connectors, such as transportation lines, which help distribute human activity across administrative boundaries. In the case of the historic city centre, it shows that administrative borders and significant natural barriers cannot limit the human consumption practices which are motivated by the functional uniqueness of the area.

The presented approach can be used for revising the administrative boundaries within and beyond the city, by reconsidering inclusion of highly connected parts, dividing disconnected ones or removing the barriers creating disconnections. This can serve as a guidance for urban planners to acknowledge natural and artificial barriers that deteriorate the use of the urban space and distort the flows of people and their practices. At the same time, urban planners can identify potential centres inside the borders of the city and on the level of agglomeration. The approach can also be implemented by city managers for better monitoring and evaluation of the quality of urban life in terms of service provision at different administrative scales.

Acknowledgements. This paper was supported in 2021 by Russian Science Foundation, project 21-77-10098 "Spatial segregation of the largest post-Soviet cities: analysis of the geography of personal activity of residents based on big data".

References

1. Aksyonov, K.: Social segregation of personal activity spaces in a posttransformation metropolis (case study of St. Petersburg). Regional Res. Russia **1** (2011)
2. Allen, G.L.: Spatial abilities, cognitive maps, and wayfinding. Wayfinding behavior: Cognitive mapping and other spatial processes 46–80 (1999)
3. Arcaute, E., Hatna, E., Ferguson, P., Youn, H., Johansson, A., Batty, M.: Constructing cities, deconstructing scaling laws. J. Royal Soc. Interface **12**(102), 20140745 (2015)
4. Cottineau, C., Finance, O., Hatna, E., Arcaute, E., Batty, M.: Defining urban clusters to detect agglomeration economies. Environ. Plann. B Urban Anal. City Sci. **46**(9), 1611–1626 (2019)
5. De Certeau, M., Mayol, P.: The Practice of Everyday Life: Living and Cooking. vol. 2. U of Minnesota Press (1998)
6. Glaeser, E.L., Kominers, S.D., Luca, M., Naik, N.: Big data and big cities: The promises and limitations of improved measures of urban life. Econ. Inq. **56**(1), 114–137 (2018)
7. Hawelka, B., Sitko, I., Beinat, E., Sobolevsky, S., Kazakopoulos, P., Ratti, C.: Geolocated Twitter as proxy for global mobility patterns. Cartogr. Geogr. Inf. Sci. **41**(3), 260–271 (2014)
8. Jones, K., Manley, D., Johnston, R., Owen, D.: Modelling residential segregation as unevenness and clustering: a multilevel modelling approach incorporating spatial dependence and tackling the map. Environ. Plann. B Urban Anal. City Sci. **45**(6), 1122–1141 (2018)
9. Langford, M., Higgs, G.: Accessibility and public service provision: evaluating the impacts of the post office network change programme in the UK. Trans. Inst. Br. Geogr. **35**(4), 585–601 (2010)
10. Lefebvre, H.: The Production of Space. Blackwell, London (1991)
11. Lynch, K.: The image of the environment. In: The Image of the City, pp. 1–13. MIT Press, Cambridge, Mass. (1960)
12. Martin, D.G.: Enacting neighborhood! Urban Geogr. **24**(5), 361–385 (2003)
13. Neal, Z.: Structural determinism in the interlocking world city network. Geogr. Anal. **44**(2), 162–170 (2012)
14. Neal, Z.: The backbone of bipartite projections: inferring relationships from coauthorship, co-sponsorship, co-attendance and other co-behaviors. Soc. Networks **39**, 84–97 (2014)
15. Nenko, A., Koniukhov, A., Petrova, M.: Areas of habitation in the city: improving urban management based on check-in data and mental mapping. In: International Conference on Electronic Governance and Open Society: Challenges in Eurasia, pp. 235–248. Springer (2018)
16. Neutens, T., Delafontaine, M., Scott, D.M., De Maeyer, P.: A GIS-based method to identify spatiotemporal gaps in public service delivery. Appl. Geogr. **32**(2), 253–264 (2012)
17. Omer, I.: Evaluating accessibility using house-level data: a spatial equity perspective. Comput. Environ. Urban Syst. **30**(3), 254–274 (2006)
18. Purcell, M.: Neighborhood activism among homeowners as a politics of space. Prof. Geogr. **53**(2), 178–194 (2001)
19. Patti, C., Frenchman, D., Pulselli, R.M., Williams, S.: Mobile landscapes: using location data from cell phones for urban analysis. Environ. Plann. B. Plann. Des. **33**(5), 727–748 (2006)
20. Ratti, C., et al.: Redrawing the map of Great Britain from a network of human interactions. PloS One **5**(12), e14,248 (2010)

21. Reardon, S.F.: Measures of ordinal segregation. In: Flückiger, Y., Reardon, S.F., Silber, J. (eds.) Occupational and residential segregation, pp. 129–155. Emerald Group Publishing Limited (2009)
22. Reed, J.S.: The heart of dixie: an essay in folk geography. Soc. Forces **54**(4), 925–939 (1976)
23. Reed, J.S., Kohls, J., Hanchette, C.: The dissolution of dixie and the changing shape of the south. Soc. Forces **69**(1), 221–233 (1990)
24. Robinson, A.: Framing corkerhill: Identity, agency, and injustice. Environ. Plann. D Soc. Space **19**(1), 81–101 (2001)
25. Schmidt, D.H.: The practices and process of neighborhood: The (re) production of Riverwest, Milwaukee. Wisconsin. Urban Geography **29**(5), 473–495 (2008)
26. Schnell, I., Yoav, B.: The sociospatial isolation of agents in everyday life spaces as an aspect of segregation. Ann. Assoc. Am. Geogr. **91**(4), 622–636 (2001)
27. Seager, W., Fraser, D.S.: Comparing physical, automatic and manual map rotation for pedestrian navigation. In: Proceedings of the SIGCHI Conference on Human Factors in Computing Systems, pp. 767–776 (2007)
28. Sobolevsky, S., Campari, R., Belyi, A., Ratti, C.: General optimization technique for high-quality community detection in complex networks. Phys. Rev. E **90**(1), 012,811 (2014)
29. Sobolevsky, S., Szell, M., Campari, R., Couronne, T., Smoreda, Z., Ratti, C.: Delineating geographical regions with networks of human interactions in an extensive set of countries. PLOS One **8**(12), e81,707 (2013)
30. Talen, E., Anselin, L.: Assessing spatial equity: an evaluation of measures of accessibility to public playgrounds. Environ Plan A **30**(4), 595–613 (1998)
31. Tsou, K.W., Hung, Y.T., Chang, Y.L.: An accessibility-based integrated measure of relative spatial equity in urban public facilities. Cities **22**(6), 424–435 (2005)
32. Zhang, T.: Decentralization, localization, and the emergence of a quasi- participatory decision-making structure in urban development in Shanghai. Int. Plan. Stud. **7**(4), 303–323 (2002)

Data Encoding for Social Media: Comparing Twitter, Reddit, and Telegram

Ivan S. Blekanov(✉) ⓘ, Nikita A. Tarasov ⓘ, Dmitry Nepiyushchikh, and Svetlana S. Bodrunova ⓘ

St. Petersburg State University, St. Petersburg, Russia
{i.blekanov,s.bodrunova}@spbu.ru, nkt.tarasov@yandex.ru,
dmitry26072011@gmail.com

Abstract. Social networking platforms have become a major source of data for most textual machine learning models. Applications of encodings in earlier language models, as well as advancements in model reuse, have opened new possibilities for case studies with limited or unsupervised data. In this paper, the authors test whether semantic similarity of large-scale data from three platforms allow for applying the same transfer-learning language models on data from various social media. For this, the authors perform a comparative case study to outline linguistic differences and measure similarity for deep neural encodings for the case data. In particular, semantic similarity is evaluated using traditional text similarity metrics, structure metrics of the corpora, and RUBERT encodings that provide general semantic characteristics of the text data in three datasets. We show that the platforms are similar in semantic terms well enough for the transfer learning models to be applied, by both linguistic metrics and semantic encodings. We also demonstrate, however, that, despite the difference in the average text length, Twitter is more similar to Reddit than to Telegram by linguistic metrics, which hints to the idea of 'platformization' of social media speech. We conclude by stating the speech factors that may lead to platform dissimilarity.

Keywords: Social network analysis · Twitter · Reddit · Telegram · Text similarity assessment · Linguistic metrics · Semantic neural encodings · RUBERT

1 Introduction

Social media play an important role in the discussions of major economic, political, and social events whether in the form of user messages or articles published by major news agencies and experts. There are a variety of platforms where opinions and discussions are common. Proper data collection and analysis is important for recognizing the socially beneficial and negative features, e.g., information spreading (helpful in informing people on emergencies such as epidemics) or spreading of fake news, among others.

The spread of social media has been also conceptualized as the 'platform society' [1] where user interactions, as well as the results of discussions, depend on platform affordances. One of the questions still unexplored is whether platform-based speech itself changes depending on the platform on which it appears. In particular, we ask whether the semantic 'profiles' of the platforms differ. Automated social media research may be divided into two types: structural analysis of user discussions and content analysis. There are a variety of studies related to the structural features such as hidden community detection or influencers detection [2] that try to relate discourse features to structural aspects of discussions [3]. However, the opposite is not always true: studies related to content typically do not utilize structural features, focusing instead on the content itself. For instance, tasks such as sentiment analysis, topic modelling, or summarization, in most cases, assess user messages without taking into account the affordances of and speech habits specific to social media platforms, while parlance on different platforms may vary significantly – which remains understudied.

In addition, data collection and data mining research has proven to be one of the major driving forces in creating robust and interpretable natural language processing models. Studies on social media are especially useful, as social networking platforms encompass a major portion of all text content on the internet. However, the question remains whether the models trained and fine-tuned for one platform may be easily used for another platform, as there are no enough studies on semantic discrepancies between social media sites on the platform level.

Comparing different social media platforms for their semantic similarity both between themselves and with traditional machine learning datasets can help economize substantially on the resources spent on machine learning and may be done in a variety of ways. General statistical approaches (length estimation, timelines of posts and comments etc.), complexity measures (a variety of linguistic measures, aimed at assessing the potential difficulty of text), and temporal features (engagement characteristics, spread of information etc.) can be used to provide a detailed comparison between different platforms. In this paper, we focus on one specific aspect of comparative modelling of user speech – platform-wide semantic similarity – via assessment of linguistic features and deep learning language models.

In particular, testing the linguistic similarities of posts and comments from different social media platforms may allow for assessing the limitations of today's most advanced transfer learning. Transfer learning [4] is a core principle of most modern language models. It allows researchers to reuse previously trained models for new sets of data by fine-tuning them. This procedure saves time, computational resources and, most importantly, requires much less training data to be used, by utilizing models trained on bigger datasets. Initial training, in this case, captures the semantic structure of the language or languages in question. Further training is required only for better domain understanding. A problem common for this approach is limited availability of datasets suitable for training models for specific social media platforms. For example, sentiment analysis data is readily available for Twitter data; supervised summarization data is common for Reddit data; etc. It is not uncommon to use a model trained for a specific social media platform for a task related to another platform [5], but the semantic discrepancy between platforms is not taken into consideration, as there is no method to assess it yet.

In this study, we start approaching the elaboration of such a methodology. We aim at measuring the large-scale platform-wide semantic differences between social networks for the purpose of statistical and language modelling applications, to find linguistic and statistical relations between posts and comments from different social media platforms in different languages. This combined approach utilizes neural encodings used for general language modelling with additional linguistic information providing more specific data characteristics in the absence of available fine-tuning data.

The remainder of this paper is organized in the following way: Sect. 2 provides a review on social media comparison methodologies previously used to measure the differences in the structure of platforms. Section 3 provides the methodology including the metrics. Section 4 provides the results and discusses their implications.

2   Previous Research

Previous platform-wide comparisons of social networking sites focused on structural aspects of user discussions, such as user connection networks, and, therefore, utilized primarily graph theory approaches [6]. We, though, will focus on content, more precisely on semantic similarity of platforms and the textual features that may influence it.

Text comparison studies encompass two main approaches: comparisons of text corpora ('corpus analysis') and similarity detection for individual texts. Corpus analysis includes such tasks as readability assessment, statistical analysis etc. Text similarity assessment is a method commonly used for plagiarism detection, general authorship detection [7], web search, and storage optimization [8].

Our approach tries to solve the problem (not tackled in prior social media research) of detecting the general semantic differences between platforms, as measured by potential impact of transfer learning model reuse for different social media platforms. We combine the ideas from text similarity studies with corpus analysis, in order to test the large-scale textual similarity of the platforms.

3   Experiment

3.1   Data

A case study was performed to detect the differences between posts and comments from various social media platforms in two languages, namely Russian and English.

For this preliminary analysis, a specific case has been chosen to study thematically identical sets of data. COVID-19 has been chosen as a topic for this comparison, as studies on this topic and data became prevalent in recent years. In this case, the choice of topics is not crucial, as it only limits the dispersion within the data, as texts on different topics can be expected to vary greatly even within the same social network.

The datasets were collected from three platforms for 2020 and 2021:

- Twitter: Posts and comments in Russian and English, 165,102 texts in total.
- Reddit: Post titles combined with post contents (if available) and comments, collected from r/worldnews and r/covid19, 195,747 texts in total.
- Telegram: Messages in Russian and English, 209,733 texts in total.

The main arguments for the choice of specific social networks are their popularity and data availability.

3.2 Metrics

3.2.1 Linguistic Measures

A few general linguistic measures, readability specific measures [9] and statistical measures were chosen to compare texts in these datasets:

- Number of sentences (N_s), number of characters (N_c) – texts with shorter average length are harder to use for most traditional models (such as topic modeling tasks) and texts which are too long can be difficult to use in deep learning language models.
- Hapax [10] – common metric in corpus analysis, indicates the amount of words only occurring once within a specific context (user message in this case). High proportions of these words may indicate a very specific language and as such may pose a difficulty for sequence to sequence tasks such as summarization.
- Number of special characters such as hashtags and reply indicators (not including punctuation) – significant amounts of special characters may negatively impact models and as such require more preprocessing.
- Punctuation – same as above, but less specific to different social networks.
- Named entities, e.g. names, companies, geographic locations – usually highly specific for topics and languages.
- ASL – average sentence length, pose the same problems as other length measures.
- ASW – average word length in syllables, pose the same problems as other length measures.
- FRE [11] – one of the most widely used readability measures, rates the text from 0 to 100 in terms of ease of understanding for specific texts.

  Formula for English language:

  $$FRE = 206.835 - 1.015 \times ASL - 84.6 \times ASW$$

  Formula for Russian language:

  $$FRE = 0.5 \times ASL + 8.4 \times ASW - 15.59$$

- SMOG – metric, shown to be outperforming FRE in specific instances [12]. Estimates the readability score based on years of education needed to understand a specific text. Formula for English language:

  $$SMOG = 1.043 \times \sqrt{(30/NS \times PW)} + 3.1291$$

  Formula for Russian language:

  $$SSMOG = 1.1 \times \sqrt{(64.6/NS \times PW)} + 0.05,$$

  where NS is the number of sentences and PW is the number of polysyllable words (3 or more syllables).

- Long words proportion (LongW) – words containing more than 3 syllables.

- Short words proportion (ShortW) – number of words with less than 3 characters.
- Functional words – structural words, not related to content (particles, pronouns etc.).
- Verbs, nouns etc. – proportions of particular part of speech words in text.
- Profane words proportion – calculated using data from profanity-filter Python library.
- Emojis – number of emojis for text, especially important for sentiment models.

Chosen set of metrics provides a diverse representation of text content from social media and allows for comparison of data across two languages.

3.2.2 Semantic Neural Encodings

Language models were used in a variety of different text analysis tasks. This made it possible to achieve state-of-the-art results for translation, summarization, question answering, and other applications. BERT – Bidirectional Encoder Representations from Transformers [13] achieves this by utilizing bidirectional encodings. Moreover, it was one of the first models to implement transfer learning as described in the introduction. This idea is particularly useful, as it can capture latent encodings for texts without the need for supervised data. RUBERT [14] is a BERT version trained on Wikipedia and news articles in Russian.

In the next section we combine these semantic neural encodings and traditional statistical and linguistic metrics to find the extent of semantic similarity between data from different social networks.

3.2.3 Similarity Assessment Measure

To measure semantic similarity of data we employ a standard cosine distance-based approach commonly used in the tasks related to text similarity assessment. To do this feature vectors for each text are averaged across individual corpora and further normalized using the min max normalization procedure. The cosine distance is calculating as follows:

$$SIM(X, Y) = \cos(\theta) = \frac{\sum x \bullet y}{\sqrt{\sum x^2} \bullet \sqrt{\sum y^2}} \tag{3.1}$$

where X and Y are feature vectors and x_i, y_i are their respective components (linguistic measures or semantic neural encodings). Values closer to 1 indicate higher degree of similarity.

3.3 Experiment Results

The metrics described in Sect. 3.2 have been used to encode the texts obtained from the three social media platforms. Table 1 shows the average values across each individual dataset with their respective standard deviations. It can be seen that the average values lie close for multiple metrics, although the deviation for some metrics indicates high instability inside each dataset.

Table 2 shows the similarity between the datasets using the formula (3.1). The table shows that linguistic similarity is especially prominent between Twitter and Reddit data.

Table 1. Mean and Standard Deviation for collected data

Metric	Twitter		Telegram		Reddit	
	Mean	SD	Mean	SD	Mean	SD
N_c	200.09	81.583	126.67	93.383	205.064	311.09
N_s	2.44	1.384	2.051	3.118	2.746	3.191
Hapax	0.768	0.134	0.881	0.18	0.75	0.212
FRE	92.7	99.112	39.59	81.32	64.279	98.366
Hashtags	4.134	3.962	0.046	0.431	0.042	0.476
Prop Nouns	5.337	4.627	2.227	8.136	1.066	2.883
ASL	16.969	10.306	7.785	10.034	13.565	8.719
ASW	5.277	1.261	1.542	4.778	1.522	1.157
SMOG	4.804	5.284	2.854	3.652	7.3	3.413
Punct	4.949	3.969	2.701	8.601	4.9	8.377
LongW	0.053	0.075	0.061	0.203	0.062	0.086
FuncW	0.078	0.096	0.054	0.106	0.203	0.105
Verb	0.057	0.077	0.053	0.105	0.198	0.092
Noun	0.275	0.144	0.430	0.409	0.188	0.128
ShortW	0.327	0.092	0.2	0.214	0.337	0.143
Profane	0.001	0.003	0.001	0.011	0.005	0.028
Emojis	0.344	1.280	0.354	2.267	0.015	0.218

Table 2. Semantic similarity of the datasets, as measured by cosine similarity (3.1) using linguistic metrics

	Twitter	Telegram	Reddit
Twitter	—	0.741	0.982
Telegram	0.741	—	0.606
Reddit	0.982	0.606	—

Additionally, each dataset was encoded using the RUBERT model, pretrained for Russian and English languages. Table 3 shows the semantic similarity using the cosine similarity measure (3.1).

Table 4 shows the cosine similarity for each dataset using both the metric vectors and the semantic neural encodings with further normalization. Tables 3 and 4 show that semantic similarity is the highest between Telegram and Reddit data.

A classification model was built to test the efficiency of the proposed metrics. XGBoost [15] was chosen as a fast and reliable classification model.

Table 3. Semantic similarity of the datasets, as measured by cosine similarity (3.1) using RUBERT encodings

	Twitter	Telegram	Reddit
Twitter	–	0.711	0.692
Telegram	0.711	–	0.92
Reddit	0.692	0.92	–

Table 4. Semantic similarity of the datasets, as measured using cosine similarity (3.1) on the combined and normalized metrics and RUBERT encodings

	Twitter	Telegram	Reddit
Twitter	–	0.719	0.716
Telegram	0.719	–	0.9
Reddit	0.716	0.9	–

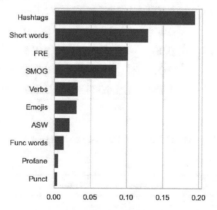

Fig. 1. Feature importance, as measured by the XGBoost classification algorithm

In this case, the classification target is a social media platform identifier. Basic feature tuning was performed using the HyperOpt library. The resulting model has achieved the accuracy of 94.8%, with the feature absolute importance figures shown on Fig. 1. Here, feature importance can serve as a measure of the most characteristic parameters, which are more likely to indicate that a text message belongs to a specific platform.

4   Conclusion

From the obtained results, we can make several conclusions.

First of all, the experiment has shown that different social media platforms can be quite similar in terms of linguistic appearance, as measured by both linguistic metrics and semantic neural encodings (using RUBERT). Thus, all the platforms look similar enough in semantic terms for transfer learning models to be applied to them without significant fine-tuning, which means that economizing on resources for transfer learning should be feasibly achievable. In particular, these similarities can be of value when transferring sentiment models (widely available for Twitter) and summarization models (common for Reddit) between these platforms.

At the same time, however, the results suggest that the platforms still differ to some extent, and the patterns of their dissimilarity by linguistic metrics and semantic neural encodings do not coincide. Thus, by linguistic metrics, Reddit is very similar to Twitter while being less similar to Telegram, which goes against the expected average text length on these platforms and hints to other platform features to be playing a role in speech divergence. Thus, the classification tests have shown that the number of hashtags, the proportion of short words, and readability metrics are different for specific platforms and are used by the classifier to distinguish between them.

On one hand, high impact of hashtag use indicates that, potentially, with good enough pre-processing, language models can be successfully transferred between platforms. This idea of heavy pre-processing coincides with methodologies and results shown in the paper introducing the Universal Sentence Encoder model [16]. On the other hand, word length and readability that also have much to do with platform dissimilarity in our data, tell that platform affordances may foster the creation of 'platformized' speech which may hinder proper use of transfer learning language models and needs to be taken into consideration.

Future work involves studying more approaches to text modelling with more linguistic parameters, more language models, and varied datasets, including juxtaposing the results for social media platforms with other types of textual data. We will focus on the textual features that lead to both similarity and differentiation of social media platforms for transfer learning.

Acknowledgments. This research has been supported in full by Russian Science Foundation, grant 21-18-00454 'Mediatized communication and public deliberation today'.

References

1. Van Dijck, J., Poell, T., De Waal, M.: The platform society: Public values in a connective world. Oxford University Press (2018)
2. Blekanov, I., Bodrunova, S.S., Akhmetov, A.: Detection of hidden communities in Twitter discussions of varying volumes. Future Internet **13**(11), 295 (2021)
3. Bodrunova, S.S., Blekanov, I., Smoliarova, A., Litvinenko, A.: Beyond left and right: real-world political polarization in Twitter discussions on inter-ethnic conflicts. Media Commun. **7**, 119–132 (2019)
4. Zhuang, F., et al.: A comprehensive survey on transfer learning. Proc. IEEE **109**, 43–76 (2020)
5. Blekanov, I.S., Tarasov, N., Bodrunova, S.S.: Transformer-based abstractive summarization for Reddit and Twitter: single posts vs. comment pools in three languages. Future Internet **14**(3), 69 (2022)
6. Guzman, J.D., Deckro, R.F., Robbins, M.J., Morris, J.F., Ballester, N.A.: An analytical comparison of social network measures. IEEE Trans. Comput. Soc. Syst. **1**(1), 35–45 (2014)

7. Koppel, M., Schler, J., Argamon, S.: Computational methods in authorship attribution. J. Am. Soc. Inform. Sci. Technol. **60**(1), 9–26 (2009)
8. Sitio, A.S.: Text message compression analysis using the LZ77 algorithm. INFOKUM **7**(1), 16–21 (2018)
9. Blinova, O., Tarasov, N.: Complexity metrics of Russian legal texts: selection, use, initial efficiency evaluation. Computational Linguistics and Intellectual Technologies (2022)
10. Faltýnek, D., Matlach, V.: Hapax Remains: authorial features of textual cohesion in authorship attribution. Preprint (2020)
11. Si, L., Callan, J.: A statistical model for scientific readability. In: Proceedings of the Tenth International Conference on Information and Knowledge Management, pp. 574–576 (2001)
12. Meade, C.D., Smith, C.F.: Readability formulas: cautions and criteria. Patient Educ. Couns. **17**(2), 153–158 (1991)
13. Tenney, I., Das, D., Pavlick, E.: BERT rediscovers the classical NLP pipeline. In: Proceedings of the 57th Annual Meeting of the Association for Computational Linguistics, pp. 4593–4601 (2019)
14. Kuratov, Y., Arkhipov, M.: Adaptation of Deep Bidirectional Multilingual Transformers for Russian Language. arXiv preprint arXiv:1905.07213 (2019)
15. Chen, T., & Guestrin, C.: Xgboost: a scalable tree boosting system. In: Proceedings of the 22nd ACM SIGKDD International Conference on Knowledge Discovery and Data Mining, pp. 785–794 (2016)
16. Cer, D., Yang, Y., et al.: Universal sentence encoder. arXiv preprint arXiv:1803.11175 (2018)

Digital Labour Platforms as New Relational Structures: Principles, Characteristics and Development Drivers

Tatiana Lukicheva[1] (iD), Kerstin Pezoldt[2] (iD), and Olesya Veredyuk[1](✉) (iD)

[1] St. Petersburg State University, St. Petersburg, Russian Federation
o.veredyuk@spbu.ru
[2] Ilmenau University of Technology, Ilmenau, Germany
Kerstin.pezoldt@tu-ilmenau.de

Abstract. Digital labour platforms (DLPs) as an innovative technology transforms not only the space of labour services relations, but also these relations as such. This determines the importance of the paper's objective, which is to attempt a generalised conceptualisation of DLPs as a new relational phenomenon. In particular, we specify and discuss the "wisdom of crowds" principle that is common across the diverse forms of DLPs. We explore the key characteristics of the new labour services relational model in terms of the way work/tasks are allocated, the configuration of roles and the bargaining power of agents. We consider the development drivers of DLPs as a new relational space for labour services. We focus particular attention on the propensity of agents, primarily young people, to use DLPs as a networking space to realise their professional, creative talent. We argue for a proactive vision of the higher school in developing students' professional skills for this new relational space and designing an enabling information and communication environment.

Keywords: Digital labour platform · Labour services relations · Labour market

1 Introduction

Digital platforms constitute, from a technological point of view, online algorithms for matching requests from buyers (users) and sellers (suppliers) for goods, services, ideas, innovations, etc., which are provided online, offline or in combination. Digital labour platforms (DLPs) are an example of a new but rapidly evolving kind of digital platforms. They are specialized in labour services.

These platforms originated in the EU countries and the US. The speed of their development has come as a surprise to researchers and policy makers [6]. Currently available figures show a trend towards further expansion of DLPs [7] not least because of the government support as in China, for example [17].

DLPs are evolving from an innovative business technology into a segment of the labour market transforming the respective relations. The approach to DLPs as a labour

A. Antonyuk and N. Basov (Eds.): NetGloW 2022, LNNS 663, pp. 123–131, 2023.
https://doi.org/10.1007/978-3-031-29408-2_9

relational structure allows seeing them as a tripartite "performer – platform – customer" model, which is different from the traditional bilateral "employee – employer" relations.

This new model operates on principles that are not always aligned with those of the traditional labour relations model. It is clear that DLPs is not an overnight phenomenon, but a long-term transformation of the world of work influenced by the macro-trend of digitalisation of the economy. The aim of this paper is to scientifically comprehend and conceptualise the phenomenon of DLPs and to provide testable hypotheses in respect to this new space of relations regarding labour services.

Next, we will discuss crowdsourcing as an operating principle of DLPs. Then we will look at DLPs as a new model of labour services relations. After that, we will describe the fundamental relational transformations resulting from DLPs' nature. Following this, we will highlight the development drivers of DLPs as a new space of labour services relations. Here we will focus on the readiness of the young generation to use DLPs as a networking space and the corresponding proactive role of higher schools. We end the paper with conclusions.

2   Crowdsourcing as DLPs' Functioning Principle

Despite the diversity of DLPs' forms, they all are based on the well-known principle of "wisdom of crowds" that large groups of people are collectively smarter than individual experts [16].

Today, the Internet and its associated technological capabilities modified this principle into crowdsourcing and crowdworking (specific for labour services) [1, 5]. The process of interactions based on this principle is organized on platforms as follows: the crowdsourcer (a firm, a group of individuals or an individual) sets a task for an indefinite number of potential participants, the crowdworker (an employee, freelancer or a separate firm), through an open offer. The crowdworker selected by the customer undertakes the task. Thus, crowdsourcing platforms as a place for crowdsourcer and crowdworker interactions act as a transaction environment for the parties. However, there are other functions – from registration to task assignment and reward – which are also performed and controlled through the platform.

There are two fundamentally different models for organizing such platforms. The first one assumes that the crowdsourcer her/himself creates her/his own platform. The second is based on the involvement of a third-party – intermediary platform. Crowdsourcing intermediary platforms help crowdsourcers to formulate tasks and requirements for their implementation. In addition, they oversee the crowdsourcing process and are responsible for the activities of the crowdworkers related to the platform [11, 12]. These platforms also assist crowdworkers by placing their labour services offers and providing related support. In general, such platforms can be seen as brokers that bring together crowdsourcers that demand certain knowledge, skills, on the one hand, and their providers (crowdworkers), on the other, by providing the infrastructure required for crowdsourcing activities. It is important to emphasize that the platform, as an intermediary, plays a key role in this relationship, as it insures access to a large amount of resources for crowdsourcers and crowdworkers.

At least two types of crowdsourcing can be distinguished: competitive (tournament) and collaboration-based. In the first case, participants choose their own tasks from the proposed set and work on them independently; in the second case, they look for other participants with whom they collaborate to solve the formulated problem [1].

The established taxonomy of DLPs is based on whether the labour services provided are location-specific or can be performed independently of location [15]. On this basis, the platforms differ in the nature of their operation, the position of the contractors, the applicable legal regulations and availability of social guarantees. If an action is to be performed at a certain place, at a certain time and only by a certain person who is then personally responsible for its performance, this is gigwork. A more detailed taxonomy in this group implies the introduction of additional criteria, such as the degree of personalization of the service, the possibilities and risks the performer should be ready to bear. In case the execution of the task is location independent, i.e. it can be done entirely over the Internet, this platform is called cloudwork. Besides, if it doesn't matter who is executing the task, and the task can be offered to an open non-specific group (crowd), this is a crowdwork. The task can be divided between group members (crowd) into smaller pieces with a fixed price for each. This is a case of "microtasking". And finally, if everyone in the group solves the same task in parallel and at the end only one result is selected and paid for, it is a case of a creative contest.

To implement the crowdsourcing principle, so-called "social media applications" are usually used. This refers to digital media and technologies that allow users to share information with each other and develop content both individually and together. Since the most common in today's digital economy are intermediary crowdsourcing DLPs, we proceed with more detailed analysis of their functioning.

3 Transformations in Labour Services Relational Model Resulting from DLPs' Nature

The expansion of DLPs is characterized by the disintegration of the labour relational model traditional for the 20th century in Western countries, where a full-time worker had a permanent employment contract directly with her/his employer. This model is gradually ceasing to dominate [8]. Defining the new labour services relational model as "non-standard" or "atypical" rejects the features of the previous model, but hardly gives a clear picture of the essence of the new reality of the world of work. Meanwhile, online interactions are shaping and changing more and more characteristics of contemporary labour services relations [6].

Researchers identify at least three essential characteristics of the labour service relations mediated by the platform. The first characteristic concerns the way the work is distributed. The competition for task performance takes place through Internet platforms. As a result, the distribution of work is moved away from the traditional (offline) infrastructure of the labour market [4].

From the point of view of the traditional labour relations, the second characteristic is that the relations existing mediated by these platforms are quasilabour relations. On the one hand, clients-customers (crowdsourcers) do not define themselves as employers. And DLP does not impose on them the usual obligations of the employer to control the appointment, performance and evaluation of the employee's work functions [2]. On the other hand, potential performers (crowdworkers) may participate in the market anonymously. They compete with each other (the geographical boundaries of such competition go far beyond the borders of a particular country) as an individual labour force for the advertised job and remuneration, providing services under the conditions of irregular working hours. The relations that arise between crowdsourcer and crowdworker go beyond the well-established labour legislation and collective bargaining agreements. The typical social environment of the former labour relations is changing. However, working through an online platform involves significant challenges for workers: income instability (including practices of price manipulation to balance supply and demand undertaken by platforms), stress, and the difficulty of maintaining a decent standard of living. The form of remuneration also varies, and it is mainly on a contractor agreement or a fee basis [9].

The third characteristic of the labour services relations mediated by the platform is the configuration of such relations that takes place between three parties: the operator of the labour platform, the client-customer and the contractor. Platform operators see themselves solely as intermediaries between the two remaining parties; they do not produce the product or provide labour services on their own. Their revenues are generated from commission fees for intermediary services [15]. The contractual relationships arise between the platform operator and the client-customer, between the client-customer and contractor, and finally between the contractor and the platform operator.

According to Schmidt, this "triangle" of relations is not balanced in terms of the ratio of market power of the three parties mentioned. The dominant player is the platform operator, who aggregates databases with information about the other two groups of participants, between which direct initial contact is impossible. Using its market power, the platform operator dictates its terms and conditions to clients-customers and contractors through the terms of the contract concluded with them, as well as during the arrangement of the order (starting with its content to the amount of payment and acceptance), for which the operator is fully responsible. This means that both the crowdworker and the crowdsourcer find themselves in an unequal position vis-à-vis the platform operator.

Depending on the platform or type of contract, the nature of competition between contractors (crowdworkers) changes. Thus, if the reward is fixed, competition between them may arise, for example, for the speed of order acceptance ("first come – first receive"). In case of other forms of remuneration, crowdworkers are competing with each other by means of price or non-price (by offering a more appropriate concept, a better idea, a more attractive artwork of the preliminary project, etc.) means. In addition, this competition is based on the principle "the winner takes it all".

4 Drivers of DLPs' Development as a New Space of Labour Services Relations

Regarding the development of DLPs, one should mention drivers related to the environment. We will start with the number of digital users and the scale effect.

As research shows [14], countries with a large digital user base, common culture, language and legislation are more likely to have an advantage in the formation and development of DLPs. At the same time, a relatively small digital client base, limited use of Internet channels for transactions, and low trust in digital technologies can be a significant limitation. To meet this challenge, digital consumer literacy and platform "adoption" should be fostered and developed.

The next driver of DLPs' development is an innovation culture open for cooperation and information exchange. Apparently, the speed of creation of such an environment can be accelerated if large companies that simultaneously implement a culture of cooperation and collaboration initiate the development of structures, processes and ecosystem management of open platforms. This could also be facilitated by the creation of innovation centres that bring together universities, laboratories, startups and large businesses.

The third driver is the technological maturity of the environment. It describes the state of technology and digital assets, including levels of connectivity and investment in next generation technologies (such as industrial internet and artificial intelligence). These will affect platform creation, growth and scope.

The fourth driver of the DLPs' development is the educational readiness of the environment, measured by the number of specialists, their level of qualification, and the diffusion of digital entrepreneurship. Science, Technology, Engineering and Mathematics (STEM), entrepreneurial and creative skills are fundamental in ensuring digital innovation. If STEM skills become an educational priority for government programs, if they are implemented by enterprises to increase the quality of their staff, then this factor of the external environment becomes a powerful resource for digitalization.

A special role in this group of development drivers is played by the propensity and competence of agents, especially young people, to use digital services as a networking space not only when looking for a job, but also to realise their professional, creative talent. As the behaviour of those just entering labour market in respect to DLPs is our special focus of interest, we devote a separate part of the paper to it, which follows this section.

Finally, the fifth driver of the DLPs' development is responsive public policy. The reason that necessitates state regulation of the market (in this case, the platform labour market) is a potentially large scale of negative consequences caused by the failure of the market mechanism to generate efficient and socially acceptable solutions. This has implications in terms of both the dynamics of quantitative growth in relations on DLPs and the quality of those relations.

Here is a brief description of the main "failures" of the DLPs and the corresponding objectives facing the state. First, low wages, the reason for which lies in ultra-high competition between employees. In this case, the objective of the state is to implement minimum standards of remuneration of labour. Secondly, discrimination of employees due to the introduction of rating systems on platforms whose algorithms are far from being always transparent [10, 13]. The objectives of the state may involve regulating the requirements for including a description of these algorithms in the user agreement. Thirdly, low entry barriers make the DLPs not only attractive to traditionally discriminated groups of workers (young people, disabled people, older people, women, mothers caring for children [3]), but often non-alternative for them. The government's objectives in this respect may include the dissemination of information about better job opportunities and facilitation of job search skills. Finally, there is a lack of information on health risks, bullying, etc. associated with working through online platforms. The state can fund relevant scientific research and inform the society about ways to minimize such risks. The state may introduce the labour standards, which, once universally applicable in the traditional market segment, will eventually be established as the de facto default norm in the DLPs as well.

The above recommendations are not based on the desirability of the DLPs but rather on the fact of their expansion as such. Thus, the state could help to reduce negative consequences of the DLPs functioning; maintain competition both between DLPs and between platform and traditional labour markets; balance interests of different groups of DLPs' participants.

5   Propensity of Youth to Use DLPs as a Networking Space: Case Study

We would like to focus particular attention on the propensity of young people – those who are only entering labour market now but will determine it in large in the nearest future – to use DLPs as a networking space to realise their professional, creative talent.

In this regard, the results of our research, based on a series of five student projects on DLPs-moderated (in particular, freelancing) work among young people in Russia may be of interest. These projects were carried out in the period 2018–2022 under the leadership of Tatiana Lukicheva within the framework of several educational bachelor degree programs (economics and human resource management) at the Faculty of Economics of St. Petersburg State University. The unique feature of such project work is that, under the guidance of a teacher-coach, students identify an interesting problem within a designated area and conduct field research (questionnaires and interviews). The following topics related to DLPs were considered: "Remote work format" (2018), "Platform employment for future highly skilled labour market participants" (2019), "Student participation in part-time employment" (2021), "Attractiveness of platform employment for undergraduate and graduate students in view of the one-year experience of the corona crisis" (2021), "Student moonlighting: freelancing or working for a company" (2022). The key research hypotheses were hypotheses about students' preference to have a desk job rather than alternative freelance; and that students are poorly aware of the characteristics of alternative to non-desk job, including freelance via DLPs.

The target group were undergraduate students or graduates of the last one or two years from both Moscow and St. Petersburg universities as well as regional universities. Students represented law, advertising and public relations, liberal arts and sciences, linguistics, economics, management, and technical specialties. Around 80 per cent of respondents had labour market experience.

About 100 replies via Google-form to questionnaires were processed once each year. Although individual projects have likely produced unrepresentative results regarding specific questions they addressed, nevertheless, the fact that students themselves formulate and analyse problems directly relevant to them and test their relevance in their environment allows such research to be considered valid. Moreover, their fresh perspective often highlights aspects of the phenomenon under study that are not visible to outside and experienced researchers.

Our analysis based on combined results of the above-mentioned student projects shows that, overall, a rather interesting general picture emerges of Russian students' attitudes towards working through DPLs (in particular, freelancing), both during and immediately after their studies. Having an opportunity to earn money, obtain additional competences, and gain work experience, combined with independent distribution of time between work, study, and leisure is a format that attracts a large part of students. Furthermore, it turns out that if there is a real choice between working on an employment contract or freelancing, the preferences of young people in higher education unambiguously shift towards working for a company (desk-job), even if only an internship is offered. Freelancing is thus regarded by most people mainly as a way of working part-time, which is attractive, but only if they have a permanent job. Although the freedom to manage oneself and one's time, the freedom to create, are important elements in the value system of young people, the value proposition of freelancing does not compare with more stable earnings and position that comes with desk-job in a company. The latter remains relevant even if it involves initially very low incomes, simple executive work at the start, a long climb up the career ladder, suppression by bureaucracy or despotism on the part of the line manager. The unregulated and uncertain social and economic status of the freelancer discussed next (the fifth driver) also has a serious impact on the non-selection of this form of labour services relations.

The concept of freelancing is familiar to almost all students, but only a small number of them are aware of the meaning behind it. In the answers to the question of what freelancing is, it is often understood as a remote job. This is mostly the understanding of those who have not had any work experience at all. If there is work experience, the definition of freelancing is clarified and it is seen as self-employment.

Lack of experience gives rise to many myths and fears among young people. There is a shortage of knowledge and specific skills on how to enter the field, compose a CV, get the first order, price the service, protect oneself against unscrupulous customers, and build effective communication with clients. There is insufficient and unstructured information for the student to understand on his or her own. This also applies to existing freelance platforms. About 60 per cent of respondents in the surveys under consideration have not heard anything about them at all. At the same time, large freelance platforms such as Freelance.ru, FreelanceJob.ru, Freelancehunt.com, Workzilla.com, Profi.ru, Weblancer.net operate in the Russian-speaking space.

Factors influencing the likelihood of students to choose freelancing platforms for work, as shown by the research under consideration, are, among others: the level of qualification requirements; condition of mandatory payment for subscription/access to the platform; security and guarantee of payment; possibility to solve disputable issues through arbitration; availability of support service; availability of information; possibility to work without experience (for example, when a new freelancer can charge below average, self-referral, etc.). The availability of information about freelance work in different sources also affects these choices.

Working as a freelancer while studying can lead to a serious conflict, as one needs to actively build up expertise in order to reach even average earnings, which is likely to take away time needed for the student to study.

Summarising the results of our studies, it becomes clear that the propensity for young professionals to use DLPs is high, but the distance to its full realisation is long. In order to engage young people in using DLPs as a space for their professional development (rather than just looking for job or performing simple tasks that do not require qualifications), the role of institutions, in particular higher schools, in developing students' professional skills for this space and designing an enabling information and communication environment should be reinforced.

6 Conclusion

DLP, as an innovative technology based on the "wisdom of crowds" principle, is transforming not only the space of labour services relations but also these relations as such.

The development of DLPs is stimulating a shift from the traditional bilateral "employee – employer" relations to a tripartite "performer – platform – customer" labour services relations. At the same time, key characteristics of the labour services relations, which deal with the way work/tasks are distributed, the configuration of roles and the bargaining power of the agents, are being transformed.

A transformation of the social environment is taking place, which manifests itself in increased opportunities for different social groups (young people, university graduates, entrepreneurs, freelancers, artists and other creative class people, young mothers, disabled people, those living in remote geographical areas) to meet their various labour needs (preferences for routine or creative activities, income or self-realisation, etc.).

The new space of labour services relations shaped by DLPs is in its formative phase. The government, employment centers, recruiting companies, businesses, and universities are increasingly involved in its creation. The prospects of DLPs, in our view, depend on their ability to generate solutions that are not only technologically effective and economically optimal, but also socially acceptable, namely, by creating opportunities for different social groups to realise their diverse labour needs.

As our research has shown, there is a demand for universities to be proactive in helping students to develop skills for professional work in the DLP-space, and in designing a specialised information and communication environment for them.

The paper contributed to the research of DLPs by providing testable hypotheses. For sure, further research of the proposed theoretical mechanisms is required.

References

1. Afuah, A., Tucci, C.L.: Crowdsourcing as a solution to distant search. Acad. Manag. Rev. **37**(3), 355–375 (2012)
2. Berg, J.-M., Furrer, M., Harmon, E., Rani, U., Silberman, M.S.: Digital Labor Platforms and the Future of Work: Toward Decent Work in the Online-World. International Labour Office (ILO), Geneva (2018)
3. EU-OSHA: Protecting workers in the online platform economy: an overview of regulatory and policy developments in the EU. European Risk Observatory discussion paper. https://osha.europa.eu/en/publications/protecting-workers-onlineplatform-economy-overview-regulatory-and-policy-developments. Accessed 15 Jan 2022
4. Gerber, C., Krzywdzinski, M.: Schöne neue arbeitswelt? Durch Crowdworking werden aufgaben global verteilt. WZB Mitteilungen **155**(3), 6–9 (2017)

5. Howe, J.: The rise of crowdsourcing. Wired Mag. **14**(6), 176–183 (2006)
6. Huws, U., Spencer, N., Syrdal, D.: Online, on call: the spread of digitally organised just-in-time working and its implications for standard employment models. N. Technol. Work. Employ. **33**(2), 113–129 (2018)
7. International Labour Organization. World Employment and Social Outlook 2021: The role of digital labour platforms in transforming the world of work. International Labour Office – Geneva: ILO (2021)
8. International Labour Organization: World Employment and Social Outlook: The Changing Nature of Jobs. ILO, Geneva (2015)
9. Kalkhake, P.: Plattformökonomie, BMAS (Hrg.) Arbeiten 4.0 Werkheft 0.1, Berlin, pp. 52–57 (2016)
10. Kullmann, M.: Platform work, algorithmic decision-making, and EU gender equality law. Int. J. Comp. Labour Law Industr. Relat. **34**(1), 1–21 (2018)
11. Mrass, V., Peters, C., Leimeister, J.M.: One for all? Managing external and internal crowds through a single platform – a case study. In: Proceedings of the 50th Hawaii International Conference on System Sciences (HICCS), pp. 4324–4333 (2017)
12. Mrass, V., Peters, C., Leimeister, J. M.: Crowdworking-Plattformen als Intermediäre und Instrumente neuer Formen der Arbeitsorganisation, Abschlussband Projekt, "Herausforderung Cloud und Crowd", Freiburg (2018)
13. Popescu, G., Petrescu, I., Sabie, O.: Labor in the platform economy: digital infrastructures, job quality, and workplace surveillance. Econ. Manag. Financ. Mark. **13**(3), 74–79 (2018)
14. Portulance Institute. Network Readiness Index (2021). https://networkreadinessindex.org. Accessed 17 Aug 2022
15. Schmidt, F.: Arbeitsmärkte in der Plattformökonomie Zur Funktionsweise und den Herausforderungen von Crowdwork und Gigwork. Berlin (2016)
16. Surowiecki, J.: The Wisdom of Crowds: Why the Many Are Smarter than the Few and How Collective Wisdom Shapes Business, Economies, Societies, and Nations. New York (2004)
17. Yao, Y.: Uberizing the legal profession? Lawyer autonomy and status in the digital legal market. Br. J. Industr. Relat. (July), 1–24 (2019)

Political and Power Networks

From the Street to the World: Situating Urban Social Networks in Multi-scalar Relations of Power

Claire Bullen[1,2(✉)] [iD]

[1] University of Tübingen, Tübingen, Germany
claire.bullen@uni-tuebingen.de
[2] University Aix-Marseille, Aix-en-Provence, France

Abstract. Marseille is often associated with ethnically-marked groups and/or impoverished segregated neighbourhoods. In this chapter, I argue that such representations can obscure aspects of social reality in and of the city. In it, I present and examine social networks of an individual who resides and works on one of the longest and arguably most heterogeneous streets in Marseille. The data has been gathered during a long-term ethnographic study of the street. It is argued that by using the street as an entry point from where to explore social change, and situating egocentric networks within time and space and multi-scalar relations of power, a deeper and more nuanced understanding of contemporary urban socio-spatial formations/interactions in Marseille comes to light, transcending binary thinking and revealing histories of uneven urban development.

Keywords: Qualitative social network analysis · Multi-scalar · Street ethnography · Marseille

1 Introduction

Marseille is a city often associated with the visible presence of 'migrants' or different ethnically-marked and racialised groups. These social 'groups' are regularly linked with marginalised, impoverished and racialised 'neighbourhoods'. Such representations can echo and reinforce reified understandings of socio-spatial relations and geographical areas that continue to prevail in migration and urban studies more generally. They may obscure more than they reveal about the complexity of urban life.

Drawing on the conceptual framework of multi-scalar city-making of Ayse Çağlar and Nina Glick Schiller [1, 2], I propose the investigation and analysis of social networks that intersect one long, highly diverse street in Marseille. The contention is that by starting from the street it is possible to situate social networks in time, space and multi-scalar connections of differential power, gaining a deeper appreciation of the shifting dynamics shaping urban life in Marseille. The material used to build this case comes from an on-going interdisciplinary study involving archival, architectural, photographic and ethnographic research.[1] For this chapter, a life history interview of a resident and entrepreneur serves to illustrate how personal networks are formed along, in and beyond particular places. It also underscores the utility of a qualitative approach to gather data concerning ego-networks.

2   Situating and Studying Marseille

With a population of around 870,000, Marseille is currently France's second largest city. On the Mediterranean shore located 700 km from Paris, the city occupies a unique, if changing, place within the national imaginary. Its reputation as an outsider is tied up with the city's late and unwilling incorporation within the French state in the 17th century and the historic intractability of the city's leaders and population to docilely follow orders from 'Paris'. The city's historical function as the major port of empire, and gateway for south-north migration has seen Marseille portrayed and researched in ways that are strongly inflected by colonial and racialising logics [3, 4]. These associations play into the city's disproportionate connection with criminality, poverty and poor governance [5]. Notwithstanding twenty years of major state intervention of urban restructuring and cultural-led regeneration projects which have contributed to the revalorisation of Marseille's image, it only needs one incident of gang-related violence or one report of corruption in local government for Marseille to be branded once again as particularly problematic, 'wicked', 'impossible', and/or crime-ridden in ways frequently linked to the visible presence of ethnically marked others [3, 5].

For many of these reasons, Marseille has incited considerable interest for urban and migration scholars [6]. Yet the force of these imaginaries poses challenges when exploring what makes the city tick. Some fine scholarship has explored the complexity of the entangled relations of Marseille (some key contributions include: [4, 51–56]). Nevertheless, dominant assumptions about social issues in Marseille shape an entrenched geography of research. Impoverished city-centre sub-districts (*quartiers*) and areas to the north of Marseille associated with high-rise housing, often referred to by the shorthand 'the northern quarters' (*les quartiers nord*), receive the lion's share of scholarly attention [6]. Within this work, questions of access to housing, displacement, and/or the everyday lives of economically impoverished, racialised groups within the neighbourhoods are the focus. These are all important topics, and I research them too. However, dynamics shaping better-off urban districts and connections between well and less-well off areas remain understudied (though see [7–9]).

In common with urban and migration studies more widely, an ethnic, or ethno-religious 'lens' (see [10] among others) is often turned on social relations. This is evident in studies on 'the Comorians', 'Maghrebin populations' or 'Muslims of Marseille'. More recently, new social categories have emerged for study including 'artists', 'activists', 'neo-Marseillais', 'gentrifiers' or 'professionals'. In many, the lives of the impoverished (and ethnicised) and higher socio-economic groups (non-ethnicised, implicitly 'white') groups are often juxtaposed (see: [8, 11, 12]).

---

[1] The six-month fieldwork has involved close collaboration with photographer and publisher Abed Abidat and architecture graduate and historian of urban policy, Amel Zerourou.

Evidently, how Marseille is discussed and researched is not set in stone; not least because the city has been undergoing considerable transformation. Some of the most visible changes are linked to the major urban restructuring programme, 'Euro-Mediterranean', involving the transformation of 480 hectares of land over 30 years, which stretches from the city-centre along the waterfront to the northern industrial area. This represents a major effort by urban decision makers to 'upscale' [13] Marseille within the international political economy. Piloted by the central state, partially financed by the EU, the redevelopment has led to new office and residential quarters on former industrial and residential land. 'Marseille is changing' (*Marseille change*) is one of the key slogans of this urban restructuring programme. Far from a linear process, in a city where attempts at top-down regeneration regularly stall, these modifications have nevertheless led to the displacement of impoverished inhabitants from city-centre neighbourhoods [14].

Another significant change to shape social dynamics in Marseille has been the election in 2020 of a left-wing and green coalition united under the name 'The Marseille Spring' (*Le Printemps Marseillais*). The new coalition replaced twenty-five years of a right-wing administration that married overtly racist politics and opportunistic clientelist management of ethno-cultural difference. The election of the new administration built on the anger and activism following the deaths of nine people caused by the collapsing of houses in the city centre in 2019. Seen as resulting from gross negligence of the previous municipality, this event brought together individuals and groups of diverse socio-economic and cultural backgrounds around vindications for the 'right to the city' [15].

These and other transformations have fed into diverse research agendas. Nevertheless, the narratives of ethnicised and racialised 'ghettos', 'gentrification', and 'globalisation' remain too easily rolled out. More complex stories of urban transformations can be elided, including relationships across ethnicity and socio-economic class, and the role of 'migrants' within gentrification processes and neo-liberal urban development.

3 Situating and Thinking from the Street

The two-kilometre street that is the focus and entry point of this study was built in the mid 19th century. At one end, south of a long dark tunnel passing under Marseille's central train station, grand Hausmannian-style housing extended the bourgeois quarters of the expanding city centre. North of the tunnel, the street linked Marseille's port and docklands to the city station. The urban fabric was built up of industrial and commercial buildings, interspersed with working class tenements.

Considerable socio-spatial differences on either side of the tunnel remain today. The southernmost, more 'central' end falls within the 1st *arrondissement* of Marseille, where organic vegetable shops and independent bars and restaurants points to visible, though not uniform, processes of 'gentrification' [14, 16]. The other side of the tunnel crosses the economically-impoverished, more 'marginalised' 3rd *arrondissement* until it reaches the new apartments, offices and hotels built as part of the Euro-Mediterranean restructuring project.

In short, Boulevard National cuts through, connects and juxtaposes socio-spatial areas produced and operated at multiple socio-spatial scales in relation to sometimes contradictory or conflicting values and meanings [2] For over a century the street has been connecting actors, places, processes involved in unequal power relations that tend to be studied apart [6]. When situated within broader structures and processes, network thinking can help untangle some of these connections and separations.

4   Networks and Analyses of Multi-scalar Relations of Power

Social network analysis is most famously associated with the seminal studies of John Barnes [17] and his examination of class formations in a Norwegian island community, and Elizabeth Bott's [18] study of conjugal relations and class in East London. The conception of social life in terms of linkages between individuals and groups was heralded as a means to transcend reified thinking about social space, while helping to break up the flow of everyday life into something manageable to study [19, 20]. Yet, since the early heyday of the 1950s, there is no consensus over either the best methods or even the definitions of the key terms [21].

According to Knox et al. [22] two major currents developed from around this time. On the one hand, sociologists working predominantly in the US developed highly mathematical methods based on graph theory. The goal was to trace and place ego-centric network linkages within social structures so as to better understand unequal social outcomes. This research offered a number of insights into social organisation within complex societies, notably Granovetter's identification of the importance of weak ties ('acquaintances' as opposed to 'friends') when accessing employment opportunities [23]. However inherent assumptions that social outcomes could be explained solely in terms of social ties often left out questions of powers. Bourdieu's notion of cultural capital was developed to address this lack [24].

The anglophone British social anthropological approach to networks is perhaps most famously associated with the so-called Manchester School and their research of urbanisation in the Copperbelt area of colonial central Africa during the 1940–1960s [25].[3] Scholars working in this tradition (including Barnes and Bott) were concerned to trace and define characteristics of social networks as a means of transcending structural-functionalist analyses then dominant in the discipline [19: 246, 26: 20]. A major contribution of this School was their attention to how urban processes were constituted within larger overlapping multi-scalar social systems and structures, shaped by colonial capitalism, rural/urban connections and tribal ties and norms. Further, rigorous attention was given to move beyond 'network' as a loose figure of speech. Considerable intellectual effort was dedicated to defining the morphological and interactional nature of people's ties (including aspects such as density, reachability, range, directedness, durability, intensity, and content of single or multi-stranded relationships [27, 28: 11–12]).

Enthusiasm for network-centred approaches waned in the following decades [22]. In anthropology, networks were seen as adding little to the discipline's long-standing focus on relationships and exchange [28]. At worst, network thinking led to calcified views of

---

[2] Conceptual clarifications of what is meant by a street and its theoretical and empirical utility will be developed in a forthcoming paper.

[3] Within the Francophone scholarship, George Balandier and colleagues were making similar points in relation to West Africa [57].

social ties, eliding change over time and individual agency [20]. Sarah Green suggests that this shift reflected a broader move from studying relations to focus on identity from the 1980s onwards (see [29: 4–5]).

This trend was rectified by the identification of a new era of 'globalisation' around the turn of the century. Network metaphors became pervasive to capture the profound sense of living in an ever-more connected world of global capitalism [30, 31; for a critical appraisal see 32]. A different approach to networks was taken within Bruno Latour's actor-network theory and Science and Technology Studies [32] with scholars exploring interconnections between humans, objects and technologies seen as coming together within horizontal, non-hierarchical, self-organised and flexible socio-spatial 'assemblages'.

Both approaches have been influential in urban research. For example, within global cities research, major cities are explored as both nodes and as enmeshed within networks of cross border trade, financial transactions, commodities and migration [33–35]. Conversely, urban scholars using an assemblage paradigm explore urban spaces as the coming together of networks, spaces and practices, focusing on the contingency of specific assemblage that produce human and non-human aspects of cities [36]. Urban assemblage researchers differ from global cities research in that they reject theory building and focus on descriptions of contingent relations within places. However, both share an interest in connections and separations *between* cities. This is often less the case in studies examining social relations *within* neighbourhoods. Showing much continuity with the classic neighbourhood and community studies from the Chicago School, attention tends to be on group formation, kinship, ethnic groups, or networks of elites, for example, *in* neighbourhoods [37], with less said about the production of geographical spaces per se [38: 269].

Network metaphors and methods have a long history in migration scholarship, deployed to capture kinship, community and personal ties and to highlight the agency of individuals in the face of structural barriers within migration systems [39, 40]. If for years the focus was on migration pathways to advanced capitalist states, transnational migration studies in the 1990s led to a paradigm shift. Work in this vein underscored myriad interconnections crossing state borders across diverse socio-spatial domains: political, economic, religious, familial...and ethno-national [41]. Yet notwithstanding decades of research, the assumption that 'ethnicity' or 'nationality' is a tie that binds - rather than being a dimension that requires empirical investigation - remains buoyant ; for example in studies of different 'national communities' in Marseille. Non-ethnicised relationships which may shape the incorporation of people within different socio-spatial systems may end up being ignored [42, 43].

Despite the 'spatial turn' hailed within the social sciences, too little attention continues to be paid to the significance of the relative position of where processes such as settlement occur [44]. For many years, urban and migration scholars Ayse Çağlar and Nina Glick Schiller [1, 2, 13] have been advocating an urban research agenda that can capture the role of migrants within urban rescaling and urban place-making, that does not essentialise migrant identities, and can draw attention to how relations of power affect the intersection of these phenomena. In their anthropological approach, 'power' can be read as the differential possibilities of social actors to influence outcomes, including when it comes to make cities, and to have the part they play recognised. Particular geographical localities serve as entry points from where to link the networks of individual actors within particular social space to multi-scalar institutional power [2: 9]. This street ethnography is strongly shaped by this approach.

5   Introducing the Methodology and Justifying the Case

The genesis of this study dates to 2014, when I began jotting informal fieldnotes as a resident of the street. A more intense period of research took place between December 2021 and February 2022, which involved: 42 semi-structured interviews, lasting 30 min to five hours, 15 portraits of different users of the street; four life history interviews. The latter were carried out with:[4]

- French woman (80) – former civil servant; widow; moved from Paris to return to Marseille in 2016 for family reasons; owner of three flats along the south end of the street.
- Moroccan man (47) – informal builder without the right to work; married, with wife and children in rural Morocco; migrated to Marseille in 2020; resides in a small flat in an insalubrious block of flats occupied by people without residency permits (*sans papiers*) along the middle stretch of the street.
- French man (56) – a manager within a registered social housing agency; Marseille born; resides outside of the city with his partner.
- Algerian woman (42) – migrated to Marseille in 2014; owner of small eatery (*snack*) in the middle section of the street; lives with family and new partner on street.

In what follows, I present data co-produced through interviews and conversations with Zahra, the Algerian woman mentioned above. Biographical data drawn from the interview is analysed to identify networks in relation to settlement in the city, provisioning everyday interactions along the street [45]. Material is triangulated with data from semi-structured interviews and material produced during participant and floating observations. This enables forms of sociality between people who may not be considered part of one's ego-centric network – for example, in the form of intermittent interactions with clients, suppliers, beggars and distant neighbours. Thus, loose connections with 'personally-known others' and with those who might seem 'strangers' [46] but who can nevertheless be thought of as 'significant others' can be explored as a constituent part of social life [see 28]. The ethnographic data is contextualised with historic and contemporary economic, political, and demographic material.

The choice to use interview data with somebody living along the middle section of the street is deliberate. This part of the street does seem to fit the dominant geography of studies of social relations in Marseille. However, by situating this material more broadly, this approach can offer new readings of people and places that are often stigmatised and reified. In what follows I sketch out material that has been gathered during participant

---

[4] Two of the life history interviews took place in a mix of Maghreb Arabic (*derija*) and French, and were subsequently transcribed and translated into French by Zerourou. In the second phase of this project, additional life history interviews are planned.

observation and the life history interview. It is presented with thick description to aid subsequent interpretation. The data is organised around three processes – migration, settlement and provisioning – and interpreted using the multi-scalar network concept. These are discussed in the last section.

6 Zahra

I first met Zahra[5] in 2015 as a customer at her over-the-counter eatery on Boulevard National. Over time, a friendly relation was forged, helped through my attempts to speak some words of Maghreb Arabic and her ready laugh. On different visits I met one of her sisters, her 15-year-old daughter and someone I later found out was her ex-husband. In late 2021, Zahra's restaurant became a meeting place for the team of researchers, where we were warmly welcomed and fed.

The interview was conducted in French and Arabic with my co-researcher Amel Zerourou. It took place in her restaurant one week day afternoon in February 2022, during the post-lunch quiet period. The restaurant remained open and Zahra stood up twice to serve clients.[6] We began by asking her to present herself and family and describe her pathway to the street.

6.1 Arrival Pathways

Zahra was born in 1979 in the east Algerian city of Annaba to Algerian parents (mother, born 1941, occupation: stay-at-home spouse; father, born 1942, occupation: construction manager, France (1962–1972) and French teacher in Algeria), the fifth of nine siblings. When in her early 20s, Zohra married a man who was eighteen years older than her. She left her family home to move in with her husband's family. They went on to have four children. As her husband did not work, Zahra obtained a job as an accountant with the Algerian police in order to provide for her children. This was condemned by her in-laws, as deemed inappropriate for a woman. They used this as grounds for criticising her morality (to the extent of questioning whether their son was the father of her children).

After ten years, Zahra demanded a divorce. She could not move back with her family. Her widowed mother now lived in a small two-room flat with one of her brothers and his family. She applied for social housing for herself and her three children. This was unsuccessful, something she attributed to her refusal of providing sexual favours for local officials – *'in the police, if you are a woman, people expect things from you'*[7] – which she considered *'haram'*[8] and not fitting with her ethical framework – *'I want to work and earn my living. Anything that is haram is not for me'*. Even if she had been able to access social housing, she feared the social stigma if she tried to raise her children as a divorced single mother in the city. This motivated her decision to move to France in

---

[5] A pseudonym.

[6] Aware that the presence of customers may have affected the content of what Zahra says, follow up interviews are planned in Zahra's house in the near future.

[7] All text in italics is translated from French from the transcribed interview.

[8] Arabic-Islamic term for religiously forbidden actions.

2014 – *'the situation wasn't good there, as I was badly paid and I didn't have a good household. I decided to come here'*.

Her choice to come to Marseille can be seen as influenced by her father's labour migration to France in the 1960s and the presence of two older sisters (59 and 51) who had moved to Marseille some years before. The migration trajectories of the different family members are at once interrelated and differently shaped by changes to mobility regimes between France and Algeria.

In the first half of the twentieth century, the French state oversaw short-term migration of 'French Algerian Muslims' from French Algeria to metropolitan France to meet labour requirements of the national economy [47]. Immediately after the Second World War, a general labour shortage meant that Algerian labour migration to (but not settlement in) France was actively encouraged. In principle, the post-war framework gave 'French Algerian Muslim' workers the same rights as French people with regards to labour or settlement. The onset of the Algerian war of independence in 1954 led to restrictions to freedom of mobility between France and Algeria and increasing surveillance and control of people from Algeria in France was undertaken by the state [47].

A new mobility regime between Algeria and France was established following Independence in 1962. Algerians could travel to France with nothing more than an identity card, giving them a special status compared to other third country nationals (there was an implicit sense that most Algerians would return to their liberated homeland). Zahra's father travelled to France under this regime. This 'favourable' period ended in the late 1960s. The French government sought to restrict an increasingly significant Algerian presence with the introduction of visas, quotas and residency permits for labour migrants and their wives and families. The severe economic recession sparked by the petrol crisis of 1973–5 and technological developments in production processes led to growing unemployment at a time of mounting racism across Europe. In France, this was particularly directed towards Algerian workers or 'Arabs' [48: 70–71]. Visas for non-labour migration were introduced in the 1980s. Formally these had few constraints attached. They were nevertheless issued under a regime of suspicion and, gradually, additional socio-economic criteria were introduced in an attempt to filter those felt to be a 'migratory risk' – young, unemployed people, with few personal resources and lacking formal qualifications [48].

Notwithstanding the increasingly harsh immigration framework and growing Islamophobia in the following three decades, numbers of Algerians seeking to cross the Mediterranean to France grew sharply from the 1990s onwards. The terrible civil war in the 1990s and lack of economic opportunities and political instability in following years contributed to rising migration aspirations. Those unable to gain a formal visa took informal routes to circumvent the formal mobility regime [40, 48]. The migration pathways of Zahra's two sisters reflect a broader feminisation of Algerian emigration to France since the 1970s, in line with international trends [39]. By 2014, 45% of all Algerian immigrants to France were women [49]. As in other national contexts, factors affecting these changes include family reunification, changes in the labour market and increases in educational levels of women.

Zahra described professional connections within the police and her role as civil servant in the police as facilitating her visa application. She also asserted that if she had not been successful, she would have come regardless – *'I just wanted to flee the country. Either I won or I lost. I just wanted to take my children from this country and this family which wasn't kind to us'*. Two of Zahra's brothers (one is in his 50s and the other is 38) migrated to Marseille in the 2020s without visas, assisted in their voyage and settlement by resources that Zahra was able to provide. The three cases illustrate what Souiah [48] describes as a growing disregard for the legal mobility framework in Algeria.

The visa Zahra obtained did not grant the right to work. After the birth of her child in a hospital in Marseille, however, she was told by a volunteer that following bilateral agreements between Algeria and France, any child born in France to parents born in Algeria prior to 1962 was automatically French. As this was the case with the baby's father, Zahra contacted her ex-partner to send or bring over documentary evidence. He applied for a visa and travelled over with the papers within three months. Her baby's French nationality appears to have supported Zahra's request for the right to work. This can take years; for Zahra it was granted within three months, giving her different opportunities for emplacement in Marseille from those of her brothers and many others she came to know in the city.

6.2 Household Emplacement and Household Ties

Zahra initially stayed with her sister who lived in the 3rd *arrondissement*, before using money she had saved up in Algeria to rent an apartment, moving onto the middle section of Boulevard National. At this point it is useful to consider some of the dynamics affecting access to housing elsewhere along the boulevard.

As noted above, the south of Boulevard National has become increasingly popular with high socio-economic groups [14] the majority of whom come from elsewhere in France or Europe.[9] This rising attractivity is linked to forms of public investment (grants for renovating the facades of buildings, a tramline at the end of the street), and more opportunities in the 'creative sectors' following initiatives like the 2013 European Capital of Culture programme [50]. More recent phenomena, such as the Covid-19 pandemic have also played their part. The latter is associated with the migration of 'Parisians' to more provincial cities, often bringing with them greater purchasing power than many Marseille residents. Many of these newcomers have been buying up or renting property in areas which had been habitually shunned by Marseille's higher social-classes. This produced important price differentials between the two parts of the street (40 m^2 at ca 700€/month, compared to 450€/month, where Zahra was living).

Renting in the formal private sector in France is bureaucratic and costly, with agencies often only accepting guarantors living in France, employment contracts and two or three months' rent in advance. This represents significant barriers for unwaged or low-waged individuals, particularly for those unfamiliar with, or not eligible for, housing benefit support. Anecdotal evidence suggests that people categorised as migrant 'others' face additional discrimination. A young Algerian couple seeking accommodation around this area were rejected on two occasions, once because the *'owner didn't rent to foreigners, as he had had many problems with them'*; once because despite their proposition to pay six months' rent in advance and having French friends acting as guarantors, the owner *'did*

---

[9] Assertion based on semi-structured interviews with traders and ethnographic observations.

not want to take the risk, because sometimes foreigners (étrangers) *say that and leave without paying their rent'.*[10] Such factors contribute to the concentrations of revenue-poor households and/or ethnically-marked individuals in certain areas of the city, such as where Zahra settled along Boulevard National. On middle stretch where Zahra and her family reside, it is easier to find properties that are rented outside of the formal rental market. In this area, the *Arabic telephone* – a French term for the rapid oral diffusion of information predominately in Maghrebin Arabic – provides an efficient means to learn about available property. Relations of trust based on shared origins, language and religion can replace the formal guarantors or the need for an employment contract.

This part of the street, marginalised from urban regeneration policies, can usefully, can usefully be considered as what Bouillon et al. [14] describe as providing 'spaces of opportunity' for those displaced from or excluded from more 'attractive' areas. They offer emplacement possibilities through relatively easy access to accommodation and opportunities for creating small businesses. However, some property owners in the area seek to profit from high returns on investment by renting out substandard property in areas where few know their rights in relation to accommodation norms and where many have a legal status that prevents them from defending their rights in courts.

In 2018, Zahra moved to a two-storey house situated in the courtyard of one of the tenements lining Boulevard National. The property was rented via an agency in the 8th *arrondissement,* the most bourgeois quarter to the south of Marseille. She paid 450€ towards the rent; the rest was paid with state benefit. Despite the high rent, the building was not fit to live in. The property owner refused to rectify this and Zahra paid for urgent repair and maintenance works herself. Through a conversation with one of her customers (the French employer of a social landlord mentioned above) she learnt that the landlord was in derogation of his rights. She was advised to withhold two months' rent to recuperate her costs. When her landlord threatened her with expulsion, this customer intervened on her behalf, attending a meeting with the landlord and threatening to pursue him through the courts. The customer is now offered free meals 'on the house'; he told me that he regularly promotes Zahra's establishment to his work colleagues.

6.3 Making a Living

Similar to many people of lower socio-economic status educated in post-colonial Algeria, Zahra had a basic command of French when she arrived. Her accountancy qualifications gained in Algeria were not recognised in France. In order to find work, she *'searched'* until she found *'people to help'*. She drew on networks of Maghrebin Arabic speakers, but clarified it was not their nationality or ethnicity that mattered but whether they were *'good people'* or not.

Within six months, (notwithstanding giving birth to her child) she had found opportunities with two Tunisian bakers, one at the northern end of Boulevard National, the other a mile further north, in the *arrondissement* close to her sister's. She worked two shifts per day: from 5am to 2pm and then helping out at the other from 3pm to 9pm. She emphasised that she was formally employed (*déclarée*); noting that she liked to do

---

[10] Personal communication, 07.10.2022.

things '*by the book*', and to pay her dues to the French state. Sometimes, she also helped one of her sisters who worked at the wholesale market in the 15ᵗʰ *arrondissement*.

Zahra compared her capacity for work with that of her sister, who was 'only' able to hold down one job. In Zahra's case, she was driven by a desire to be able to look after her family in Marseille and Algeria. She also had considerable business acumen and expanding social networks. Three years after arriving in France, Zahra was able to purchase the commercial lease of a small 'kiosk', a twelve-metre unit along Boulevard National, with a few tables and chairs set out on the pavement. From here, she sold hot and cold sandwiches, Algerian and Tunisian pastries and bread, mint tea, coffee and cold drinks over a counter, sometimes assisted by one of her sisters. Zahra named her business after her oldest daughter, as recognition for her on-going support since arriving in France (looking after her siblings, friendship and emotional support, translation with French etc.). Guidance and a financial loan were also offered by the couple who ran the first bakery, with whom she had developed a strong relationship of confidence (she ran the bakery in their absence, when they returned to Tunisia, and said she considered them like family).

Two years later Zahra saw that a 50m² commercial unit next to her place of residence became vacant. Through her personal networks she learnt the owner was '*an Arab from Annaba*'. She was able to obtain his telephone number via the Marseille-born owner of a call shop, internet café and international money transfer service along the street. Despite the spatial proximity of their two businesses and despite both being 'of Maghrebin origin', the person who gave her the number was not a personally-known other. From my observations, their distant relationship seems to embody the social differentiation and social distance that often structures relations between people born in France of Algerian, Moroccan or Tunisian origin (described by one Algerian research participant as '*the children of France*'), and more recent arrivals from the Maghreb in the area. The latter are often careful to maintain distance from the former who no longer speak 'proper' Arabic, who can be considered less polite and respectful – too 'street' – and possibly involved in amoral and/or illicit behaviour.

In her transaction with the commercial lease owner, Zahra's gender and her social capacities played in her favour. She told how he told her: '*Listen, my daughter, I've had several offers, but they were all men. As you are the only woman, I'm going to help you a little and I'm going to leave it to you, and you can pay in instalments*'. Zahra's gender, along with her entrepreneurial drive and motivation also helped her in her relationship with the landlord of the property. In this instance, the owner was someone she sometimes described as a 'Jewish French person' (*un Français juif*), sometimes by his name. He lived in Marseille, outside of the neighbourhood. When Zahra first contacted him, he told her that all he was looking for was somebody who would pay the rent regularly and who was 'conscientious/hard-working' (*sérieux*). Prior to Zahra's contract, he had rented out the property to young men who had run a shisha café almost exclusively used by men, which had closed after less than a year. He agreed to rent the property to Zahra (for 850€/month). Additionally, he contributed to the costs of the renovations Zahra carried out to transform the shisha café into a family eatery. In this relation, conscientiousness rather than ethnicity or race was significant. Conversely, in conversations with another property owner down the street, I was told that they preferred not to rent their property to *'Arabs'* because of their fear of problems and lack of payment.

There are a huge number of social relations that intersect her business, far too many to go into much detail here, including: the participation of her new husband and partner and family members in running the business and providing childcare; relations with local and transnational suppliers; and connections with local and national administrations as tax payer. I want to finish, however, with a spotlight on her customer relations, for what this can tell us about urbanity and change within the neighbourhood.

Zahra describes most of her clients as 'people from the neighbourhood' (*les gens du quartier*) who were distinguished in her reply from people who work in the neighbourhood: *'the man* (le monsieur) *from Logirem and his colleagues; teachers from the primary school'*. Months of participant observation as a regular customer allows me to pad out this category 'from the neighbourhood'. In the morning customers included those grabbing a coffee or tea, often single men or two men who take coffee or mint tea on the pavement, perhaps with a cigarette, often interacting with personally-known others, with a greeting in Arabic: *'salam alaikum'*, which could end there or transform into a longer conversation about the family and business. Throughout the day, some older men could spend an hour or two in the café over a mint tea, exchanging informal conversation with Zahra; some were significant others, who Zahra recognised but did not know well; others she considered like a member of her family, such as an elderly man who came from her home town in Algeria. The main language spoken in the restaurant is Maghreb Arabic. Lunch time customers included manual labourers, as is evident from their paint-splattered jeans or tracksuits and employees from neighbouring businesses, including the chemists and social housing agency. On occasions I have seen people involved in arts or social associations eating over a working lunch, as I have done too. In the afternoon and early evening, particularly on a Friday, parents in the neighbourhood passed to pick up portions of take-away couscous for their families; single men eat their supper in front of a telephone. On a number of evenings, I have seen four or five men use the place as a meeting room for what seems to be a Comoran diaspora association. They did not purchase anything. For Zahra, the restaurant is a place where people feel welcome. She greets people with a smile and is willing to take time to talk to people, providing a space for refreshment and sociability, *'even if they can't pay'*. On different occasions, I saw her serve food to young men known as being 'without papers' who lived in a squat along the street. Another time, I came in and Zahra patiently was listening to an old woman with clear mental health issues and a predilection for invoking the Virgin Mary at the end of each sentence. The welcoming and homely atmosphere explains why relations here often go beyond single-strand exchanges between trader and customer, to multi-strand relations based on mutual aid, friendship and kindness, such as the relationship I have with Zahra.

Often there is a distinction in Zahra's categorisations of people 'from the quarter' and those she categorises as 'French'. The latter – a category that includes me, even though I am British and live in the neighbourhood – comprises people who live in the vicinity of her eatery who look 'white European' and those people who come to work around the street during the daytime. Our enquires confirm that the majority live in the south-central neighbourhoods or commute to work from outside of Marseille. While still the minority, the presence of such 'French' people in this 'segregated', 'Maghreb' part of the city has grown in the last years, as the 3rd *arrondissement* is now the last remaining neighbourhood close to the city centre that remains relatively accessible. Zahra's eatery is starting to become known with people of such socio-economic and ethno-cultural backgrounds. The presence of such clientele indicates that Zahra's café might contribute to a budding 'gentrification' of what remains one of the last relatively affordable areas of Marseille; something that is likely to escalate if the tram-line that has been evoked is ever built.

7 Discussion and Concluding Thoughts

Much can be said and interpreted from the above thick description of Zahra's egocentric ties which currently intersect Boulevard National. I begin this discussion by noting that Zahra generally defined her own relations and life plans in terms of kinship ties. She regularly categorises her life choices and move to Marseille in terms of being able to help her children and her extended family (*'I love my family very much. I came here to work and help my family'*). She describes how she sees her family *'all the time'*, by which she means, at her house – an epicentre for the Marseille-based kin members, – in the restaurant or via social media. Throughout her life, Zahra's family networks are clearly a source of mutual love, assistance, support. Since her move to Marseille, the loss of her parents and because of her personality and drive, Zahra describes how she has become increasingly central within these kinship interactions.

It is important to give weight to these social relations that have so much meaning and value for Zahra. The data-rich descriptions above clearly demonstrate the signification of language and shared sense of identity based on country of origin and religion for Zahra's arrival, settlement and provisioning pathways in Marseille. Yet, as this ethnographically-embedded, life-history network analysis shows, in her roles as mother, sister, wife, and also as a tenant, entrepreneur and tax payer, Zahra's primary and secondary relations in Marseille have expanded beyond her immediate kin or national or regional grouping. To cite just a few: her links with the customer from Logirem; her French Jewish landlord and customers, such as myself. The above account also demonstrates the significance of categorically-known others in Zahra's future emplacement in Marseille, including the person who helped her find out that her son could gain French nationality, and the owner of the internet café. The point is that if we begin by assuming that kinship and ethnicity is the primary factor influencing urban interactions then diverse connections and different forms of incorporation into Marseille's urban systems can be lost. Moreover, analyses that focus solely on such ties can ignore power relations that intersect social structures and institutions operating across different scales, and the significance that place can make in forms of urban incorporation [10, 44].

A geographically-situated, life-history approach to network analysis allows for the interpretation of egocentric ties in ways that speak to broader issues of power and inequality, and can depict how these change over space-time. Thinking in these terms, Zahra's decision to migrate to France and the 3rd *arrondissement* of Marseille can be interpreted as the result of a complex chain of decision-making situated within wider complex, often contradictory, sometimes overlapping dynamics that link this neighbourhood to the city, the French state and across national borders to Annaba. These include: tensions with her former in-laws; transforming gender norms and practises both sides of the Mediterranean; lack of economic opportunities in Annaba for her and her family; Zahra's knowledge about opportunities for living and working in France linked to previous family migration; shifting colonial and post-colonial relations between France and Algeria and urban regeneration policies and racially-inflected modes of capital accumulation within Marseille.

By considering Zahra's incorporation in the 3rd arrondissement in these historically-situated and relational terms, it is possible to draw out the significance of living in an impoverished part of the city, where poverty, discrimination and racism structure everyday interactions. Yet, the thick descriptions of Zahra's multiple networks also allow us to gain sense of her agency, as she drew upon her own economic, social and cultural resources to navigate and to shape opportunities and interactions in the 3rd *arrondissement* of Marseille and beyond. This includes her business acumen and a strong work ethic, a deep Islamic faith and a profound affection and driving motivation to look after her family both sides of the Mediterranean Sea, and a strong personal capacity for sociability and kindness to others. In different ways, all of these currently contribute to the production of social spaces along and beyond Boulevard National.

To sum up, this chapter seeks to contribute to empirical investigations of urban transformations in Marseille that do not take for granted either socio-spatial identities or a particular geographical scale. A methodological case is being made here for the value of ethnographically-rich, biographical network analysis as a way of shedding light on the complexity and forces inherent in city-making processes shaped in and across political, economic and socio-cultural domains occurring across a multiplicity of scales [1]. I have focused on the egocentric network of one individual. As the research progresses, the focus of discussion will be widened. Networks of different users of Boulevard National will be examined comparatively. The goal is to deepen understandings of differential opportunity structures along and beyond the street, shaped by and shaping Marseille's actual transformation, and the city's shifting place in the world.

Acknowledgements. My profound thanks go to Zahra for her welcome, cooking and the time she has given to this research. Abid Abidat and Amel Zerourou have been exemplary companions during this exploration of Boulevard National over the last months. I am very grateful to all those who have taken the time to read draft versions of this, and their feedback have significantly improved it, including: Heather Bullen, Murial Girard, Nina Glick Schiller, Franck Lamot, Karolin Mattes and friends and colleagues at the Unkut research network in Tübingen. Suggestions from the editorial committee and peer reviewers have helped make this stronger. The current phase of the project has been funded by the Federal Ministry of Education and Research (BMBF) and the Baden-Württemberg Ministry of Science as part of the Excellence Strategy of the German Federal and State Governments.

References

1. Çağlar, A., Glick Schiller, N.: Relational multiscalar analysis: a comparative approach to migrants within city-making processes. Geogr. Rev. **111**(2), 206–232 (2021)
2. Çağlar, A., Glick Schiller, N.: Migrants and City-Making: Dispossession, Displacement, and Urban Regeneration. Duke University Press, Durham (2018)
3. Hewitt, N.: Wicked city. The Many Cultures of Marseille. C. Hurst & Co, Glasgow (2019)
4. Roncayolo, M.: Les grammaires d'une ville. Essai sur la genèse des structures urbaines à Marseille. Éditions de l'École des hautes études en sciences sociales, Paris (1996)
5. Samson, M.: Marseille enprocès. La véritable histoire de la délinquance marseillaise. Cahiers libres, Paris Marseille: Éditions la Découverte Éditions Wildproject (2017)
6. Zalio, P.: Urbanités marseillaises. Marseille, terrain des Sciences Sociales. Enquête, no. 4 (November), 191–210 (1996)
7. Dario, J., Dorier, E.: La rue et l'espace public face à la fermeture résidentielle. Méditerranée. Revue géographique des pays méditerranéens/J. Mediterranean Geogr. **134**(June) (2022)
8. Géa, J., Gasquet-Cyrus, M.: Marseille: entre gentrification et ségrégation langagière. Maison des sciences de l'homme, Paris (2017)
9. Maurice, M., Deloménie, D.: Mode de Vie et Espaces Sociaux: Processus d'Urbanisation et Différenciation Sociale dans Deux Zones Urbaines de Marseille. La Haye: Mouton, Paris (1976)
10. Glick Schiller, N., Çaglar, A., Guldbrandsen, T.C.: Beyond the ethnic lens. Locality, globality, and born-again incorporation. Am. Ethnol. **33**(4), 612–33 (2006)
11. Lorcerie, F., Geisser, V.: Les Marseillais Musulmans. Open Society Foundations, New York (2011)
12. Manry, V.: Belsunce 2001: Chronique d'un Cosmopolitisme Annoncé? Marseille. Probing Clichés/Derrière Les Façades **13**, 136–145 (2002)
13. Schiller, N.G., Simsek-Caglar, A. (eds.): Locating Migration: Rescaling Cities and Migrants. Cornell University Press, Ithaca (2011)
14. Bouillon, F., Baby-Collin, V., Deboulet, A.: Ville Ordinaire, Citadins Précaires. Transition Ou Disparition Programmée Des Quartiers-Tremplins?' 220089072 (CN14 02). Centre Norbert Elias (CNRS-UMR 8562): Ministère de l'écologie, du développement durable et de l'énergie, Ministère du logement, de légalité des territoires et de la ruralité, Plan urbanisme, construction architecture (2017)
15. Peraldi, M., Samson, M.: Marseille en résistances. Fin de règnes et luttes urbaines. Cahiers libres. Paris: la Découverte. (2020)
16. Escobar, D.M.: Le processus de gentrification rend-il compte des dynamiques de peuplement des quartiers centraux de Marseille? Langage et societe **162**(4), 47–51 (2017)
17. Barnes, J.A.: Class and committees in a Norwegian Island parish. Hum. Relat. **7**(1), 39–58 (1954)
18. Bott, E.: Family and Social Network. Roles, Norms and External Relationships in Ordinary Urban Families. 2nd edn. Tavistock Publications, London (1971)
19. Hannerz, U.: Exploring the City. Inquiries Toward an Urban Anthropology. Columbia University Press, New York (1980)
20. Sanjek, R.: What is network analysis, and what is it good for?. Rev. Anthropol. **1**(4), 588–597 (1974)
21. Fuhse, J.A.: Theorizing social networks. The relational sociology of and around Harrison White. Int. Rev. Sociol. **25**(1), 15–44 (2015)
22. Knox, H., Savage, M., Harvey, P.: Social Networks and Spatial Relations: Networks as Method, Metaphor and Form (2005)
23. Granovetter, M.: The strength of weak ties. Am. J. Sociol. **78**(6), 1360–1380 (1973)

24. Bottero, W., Crossley, N.: Worlds, fields and networks. Becker, Bourdieu and the structures of social relations. Cult. Sociol. **5**(1), 99–119 (2011)
25. Evens, T.M.S., Handelman, D. (eds.): The Manchester School. Practice and Ethnographic Praxis in Anthropology. Berghahn Books, New York (2006)
26. Mitchell, J.C.: Cities, Society, and Social Perception: A Central African Perspective. Oxford [Oxfordshire]: New York: Clarendon Press; Oxford University Press (1987)
27. Gluckman, M.: Custom and Conflict in Africa. Blackwell, Oxford, Cambridge (1956)
28. Boissevain, J.: The place of non-groups in the social sciences. Man **3**(4), 542 (1968)
29. Holbraad, M., et al.: What is analysis? Soc. Anal. **62**(1), 1–30 (2018)
30. Castells, M.: Local and global: cities in the network society. Tijdschr. Econ. Soc. Geogr. **93**(5), 548–558 (2002)
31. Castells, M.: The Rise of the Network Society, 2nd edn. Wiley, Chichester, Malden (2010)
32. Sheppard, E.: The spaces and times of globalization: place, scale, networks, and positionality. Econ. Geogr. **78**(3), 307–330 (2002)
33. Friedmann, J., Wolff, G.: World city formation. An agenda for research and action. Int. J. Urban Reg. Res. **6**(3), 309–44 (1982)
34. Sassen, S.: The Global City. Princeton University Press, New York, London, Tokyo (1991)
35. Sassen, S.: The global city: introducing a concept. Brown J. World Aff. **11**(2), 27–43 (2005)
36. Farías, I., Bender, T.: Urban Assemblages: How Actor-Network Theory Changes Urban Studies. Questioning Cities. Routledge, London, New York (2010)
37. Neal, Z.P.: The Connected City: How Networks Are Shaping the Modern Metropolis, 1st edn. Metropolis and Modern Life. Routledge, New York (2013)
38. Blokland, T.: Bricks, mortar, memories: neighbourhood and networks in collective acts of remembering. Int. J. Urban Reg. Res. **25**(2), 268–283 (2001)
39. Boyd, M.: Family and personal networks in international migration: recent developments and new agendas. Int. Migr. Rev. **23**(3), 638–70 (1989)
40. D'Angelo, A.: The networked refugee: the role of transnational networks in the journeys across the Mediterranean. Glob. Netw. **21**(3), 487–499 (2021)
41. Glick Schiller, N., Basch, L., Szanton Blanc, C.: From immigrant to transmigrant: theorizing transnational migration. Anthropol. Q. **68**(1), 48 (1995)
42. Dahinden, J.: A plea for the 'de-migranticization' of research on migration and integration. Ethn. Racial Stud. **39**(13), 2207–2225 (2016)
43. Glick Schiller, N.: Transnational social fields and imperialism. Bringing a theory of power to Transnational Studies. Anthropol. Theory **5**(4), 439–61 (2005)
44. Ryan, L.: Differentiated embedding. Polish migrants in London negotiating belonging over time. J. Ethn. Migr. Stud. **44**(2), 233–51 (2018)
45. Brandhorst, R., Krzyzowski, L.: Biographical reconstructive network analysis (BRNA). A life historical approach in social network analysis of older migrants in Australia. Forum Qualitative Sozialforschung/Forum Qual. Soc. Res. **23**(1) (2022)
46. Lofland, L.H.: A World of Strangers: Order and Action in Urban Public Policy. Waveland Press, Prospect Heights, Ill. (1985)
47. Témime, E.: La politique francaise a l'egard de la migration algerienne. Le poids de la colonisation. Le Mouvement social **188**(July), 77 (1999)
48. Souiah, F.: My visa application was denied, "I decided to go anyway": interpreting, experiencing, and contesting visa policies and the (im)mobility regime in Algeria. Migr. Soc. **2**(1), 68–80 (2019)
49. Kassar, H., Marzouk, D., Anwar, W.A., Lakhoua, C., Hemminki, K., Khyatti, M.: Emigration flows from North Africa to Europe. Eur. J. Pub. Health **24**(1), 2–5 (2014)
50. Bullen, C.: Gentrified, Euro-Mediterranean, Arabic? Situating Mediterranean locations along a street in Marseille. In: Rommel, C., Viscomi, J.J. (eds.) Locating the Mediterranean: Connections and Separations across Space and Time, pp. 103–27. Helsinki University Press (2022)

51. Fournier, P., Mazzela, S. (eds.): Marseille, entre Ville et Ports. Les Destins de la Rue de la République. La Découverte, Paris (2004)
52. Mattina, C.: Clientélismes Urbains: Gouvernement et Hégémonie Politique à Marseille. Académique. Presses de Sciences Po (2016)
53. Peraldi, M.: Marseille. Réseaux migrants transfrontaliers, place marchande et économie de bazar. Cultures & Conflits, 33–34 (May) (1999)
54. Peraldi, M. (ed.): Cabas et containers. Activités marchandes informelles et réseaux migrants transfrontaliers. Maisonneuve et Larose, Paris. (2001)
55. Tarrius, A.: L'entrée dans la ville: Migrations maghrébines et recompositions des tissus urbains à Tunis et à Marseille. Revue Européenne de Migrations Internationales 3(1–2), 131–148 (1987)
56. Tarrius, A.: Les fourmis d'Europe. Migrants riches, migrants pauvres et nouvelles villes internationales. Éditions L'Harmattan, Paris (1992)
57. Balandier, G.: Sociologie actuelle de l'Afrique Noire. Dynamique sociale en Afrique Centrale. Presses Universitaires de France (1982)

Assessing Electoral Potential by Interest Groups in Social Media in the Context of Russian Political System

Darja Judina[1]([✉]) [iD] and Olga Tsybina[2] [iD]

[1] Center for Sociological and Internet Research, St. Petersburg State University, St. Petersburg, Russia
d.yudina@spbu.ru
[2] St. Petersburg, Russia

Abstract. Social media remains a valuable source of data for studying the electoral potential of parties and politicians, compensating to some extent for the shortcomings of electoral surveys. In this paper, we show how the assessment of interest groups supporting the main Russian political parties on social networking sites can be an indicator of the electoral potential of these parties. We measured the quantitative characteristics of the main Russian parties' online communities and the quantitative and qualitative characteristics of interest groups seeking support in each community. These characteristics were complemented by the network analysis of online communication between these communities. The parties with the largest number of supporters were Navalny Team and the Communist Party of the Russian Federation. An analysis of interest groups showed that the themes and goals of these communities correspond to the ideology and political actions of both parties. Network analysis revealed what kind of social capital (bonding or bridging) interest groups are more likely to make up. The research contributes to methods for assessing the electoral potential of political parties through showing how it is possible to identify specific interest groups sympathetic to parties in social media.

Keywords: Social media · Elections · Political parties · Interest groups · Russian politics · Network analysis

1 Introduction

Practice shows that public opinion polls may have low accuracy in the assessment of the electoral potential of political actors [28, 32]. The reasons for that being incorrect initial formation of the sample, the spiral of silence effects [30] and a large proportion of survey nonresponse leading to sample bias [12]. In Russian political system, another problem is the discrepancy between the existing set of political parties and the real interests of citizens [46].

O. Tsybina—Independent Researcher

© The Author(s), under exclusive license to Springer Nature Switzerland AG 2023
A. Antonyuk and N. Basov (Eds.): NetGloW 2022, LNNS 663, pp. 152–170, 2023.
https://doi.org/10.1007/978-3-031-29408-2_11

Thus, public opinion polls have three sources of distortion: non-representativeness of the sample, inconsistency between the list of parties and the interests of voters, and incorrect assessment of voters' readiness to participate in elections. In our study, we propose an alternative way to measure citizens' interest in political parties by analyzing their official pages on the social networking site VKontakte. In addition to quantitative estimates of the audiences of online communities, which are already found in various other works, we analyzed a set of interest groups that are looking for support among the online audience of Russian parties.

The method we propose has limitations in that the data of party communities on SNS (social networking sites) is not able to reflect the entire spectrum of interest groups that support these parties, and quantitative online indicators may not correspond to the number of real offline voters. However, the method can be used to assess the survey sampling limitations, as well as to identify potential groups of party supporters.

2   Methods for Measuring the Electoral Potential of Political Parties in Social Media and Their Limitations

Social media has become an influential tool in political struggles, both in democratic [1] and authoritarian countries [54]. Over the past ten years, a large number of studies have appeared that compare the popularity of political parties or politicians on social networking sites (SNS) with their results in elections [7, 26]. Some of these studies show the effectiveness of this approach in predicting electoral results [48, 51].

One of the main assumptions of these comparative studies is that various network indicators: number of subscribers to the respective political communities or personal pages, number of posts/tweets in support of a particular candidate or parties, number of likes, reposts or comments under these posts – all this reflects the electoral potential of a party or politician. Electoral potential, or the set of potential voters, is understood as all citizens who have the right to vote [35].

This method has several disadvantages and limitations. First, there is no guarantee that a subscriber to a political party or a politician on SNS has the right to vote (he/she may be a minor or citizen of another state). Secondly, there are significant differences between users of SNS and real voters in terms of socio-demographic characteristics (primarily in terms of age). There are more young people among the users of social media: they go to elections less often and have different political preferences from other age groups [37]. Thirdly, the assessment of the political support of social media users can be distorted by the activity of the so-called social bots [24]. This term refers to both automated accounts and real people whose purpose is to influence online public and SNS algorithms. Despite the noted shortcomings, we believe that data from SNS represent a resource for studying the electoral potential of parties, at least in relation to social groups that manifest themselves in social media.

When applied to the study of political parties in social media, an approach that measures the proximity between different political actors is used. An example of this is research on the links between politicians and political parties with nationalist movements [13, 40]. These works show how, by studying various elements of Internet communication (hashtags, subscriptions to groups or pages, general topics of posts), one can identify support groups for politicians or parties.

In our research, we use an approach to measuring the electoral potential which combines the study of groups that support parties and their quantitative comparison Thus, we single out groups of supporters of the main Russian political parties and estimate the size of their audience in social media.

3   Interest Groups and Political Parties

According to one of the dominant theories, interest groups are seen as the main elements influencing what decisions a party makes, from nominating candidates for elections to promoting laws [5, 6, 22]. The general idea is that interest groups are organizations that have political goals and necessary resources to finance political parties, as well as mobilize voters to vote for the desired party. Empirical studies show that political parties are supported by an alliance of interest groups [25, 47]. Hence, interest groups can be considered proto-voters who are capable of mobilizing the masses of voters. This capability of interest groups is in demand both in democratic systems and electoral autocracies such as Russia today [19].

There is no definition of an interest group that would be considered conventional in political research. Baumgartner and Leach counted ten different types of definitions of the term [3]. Their analysis of the definitions has the following conclusion: there is no single correct definition – its choice depends on the design of the study. At the same time, the researchers emphasize that a specific definition has its own assumptions and limitations, which must be reflected in the research results.

We have considered the suggested recommendation and defined the term "interest group" based on the following premises. The first premise arises from the way we discover interest groups. We start from the assumption that the pages of political parties in social media are viewed by various associations of people as platforms for disseminating information about situations that concern them, as well as asking for particular help and assistance. For example, different groups of activists (animal rights activists, historic preservation activists, etc.) may consider online communities of political parties as a means of communication with members of the party, or as a media, whose audience might be interested in supporting these groups. The second premise is that interest groups are viewed as a potentially useful resource for a political party. Since we are limited to analyzing data from the SNS, we use the size of the group's audience as an indicator of available resources. The logic is as follows: the more users of social media consume the content of the interest group, the more potential voters the party will gather if it supports their interest. Based on the listed premises, we have defined interest group as an association of users of SNS, which has its clearly defined political, economic and/or other goals, the achievement of which requires political participation. By political participation, we mean voluntary collective activity that influences the policies and actions of government authorities [50].

4   Party System and Elections in Russia

Russian party system is part of a non-democratic regime [17, 19], therefore, its structure is determined not so much by the ideological attitudes of political parties, as by their proximity to one of the opposing groups: the ruling elite and the opposition. White proposed a classification of Russian parties, taking into account this feature of the Russian political system [52]. He suggested five categories: the ruling party (United Russia), pro-Kremlin parties (Liberal Democratic Party of Russia and Just Russia), semi-oppositional parties (the Communist Party and Yabloko) and principal opposition parties (Parnas, the Other Russia and the Left Front). The roles of the main actors in Russian party system are described in detail by Gel'man [16].

The classification of parties proposed by White did not lose its relevance up to 2021, when a series of political events changed the political system: the destruction of Navalny's Staff and the entry of a new party (New People) into the State Duma. However, in our opinion, it requires two elaborations.

The first elaboration concerns the phenomenon of spoiler parties. Spoiler parties are a tool of electoral manipulation: formally registered, they do not conduct any political activity, except participation in elections. Their purpose is to divert votes from active parties with similar name and/or declared ideology [18]. Since these parties do not conduct real political work, we do not consider them in our study.

The second refinement of White's classification refers to organizations which are the antipode to spoiler parties. These are active political movements, not officially registered as parties. Listing the principal opposition movements, White did not consider the fact that they were not formally registered. The lack of a formal status does not allow political movements to participate in elections. In our study, we look at representatives of the principal opposition who actively engaged in politics in the period of 2018–2019, but did not have the official registration as parties.

Elections in Russia are a source of instability for the ruling elite. On the one hand, the elite needs elections as a source of legitimacy [19], therefore they cannot cancel them, but resort to various tools to strengthen control over the elections (such as refusal to register opposition candidates [20], forced mobilization of controlled voters [23]). On the other hand, there is a need to avoid large-scale falsifications which can lead to mass protests, as it was after the Duma elections in 2012 [45]. It opens up opportunities for the opposition to mobilize supporters to participate in protest voting [49].

Despite the undemocratic structure of the party system and elections in Russia, election results are (with certain limitations) an indicator of citizens' support for a particular political force.

5 Aim and Research Questions

The purpose of our study is to assess the electoral potential of the main Russian parties which is created in social media. Since the electoral potential is characterized by the number of potential voters, as well as by the set of interest groups that see the party as a tool for achieving their interests, the stated goal assumes answers to two questions:

Q1. What is the size of online audience of the main Russian political parties in social media?

Q2. Which interest groups seek support from parties or their audience in social media?

Interest groups also have their online audience, which can be considered as potential voters of the party whose online community a group has chosen to draw attention to their need or problem. Therefore, in a comparative assessment of the electoral potential of parties, one can also use the total audience of interest groups. Hence, the third research question:

Q3. What is the size of online audience of interest groups seeking public support in the social media communities of each of the main Russian political parties?

Another issue that cannot be ignored is related to party ideologies and their historical popularity. It is logical to assume that the types of interest groups seeking support among the audiences of a particular party are ideologically close to that party. Whether our assumption is consistent with the real picture, we can find out from the answer to the fourth research question:

Q4. Are interest groups seeking support from online party communities close to them in ideology and agenda?

In the analysis of the online interaction between interest groups and party audiences, an important element is the structure of this online communication. Its consideration makes it possible to understand how active interest groups are, as well as on which parties their interest is focused, and to what extent. Accordingly, the fifth research question is:

Q5. What is the structure of online communication between parties and interest groups?

We have combined the descriptions of the methods and results in the relevant subsections of the Results chapter. Such a structure in our case is more logical, since almost every research question requires its own separate method.

6 Sampling

In our sample of Russian parties, we selected: United Russia – the ruling party; Liberal Democratic Party (LDPR) and Just Russia – the pro-Kremlin parties; the Communist Party and Yabloko – semi-oppositional parties; Navalny Team and Libertarian Party – the principal opposition parties. Navalny Team (party Russia of the Future) and the Libertarian Party are unregistered, but most active political associations that have offices in different Russian regions [31, 42] and hold mass political events during the period of study [36, 38, 39]. These principal opposition parties, unlike the rest of our sample, were not mentioned by White.

Official communities on VKontakte (the most visited SNS in Russia) were chosen as online platforms of the parties [6]. VKontakte can be seen as the Russian counterpart to Facebook [53]. Posts and comments for the period from March 19, 2018 (right after the presidential election) to May 2, 2019 were downloaded from the official communities. After that, using regular expressions, all links to communities on VKontakte that were left in the comments under the posts of the parties' communities were automatically selected.

We labeled the communities that we found in the comments (see Appendix), then discarded all that did not fall under the category of "interest groups". A total of 1,190 unique groups were labelled, of which 214 groups fell into the category of "interest groups". From the interest groups communities, we downloaded posts with statistics for the studied period and traced the size of the audience who saw each post (download date – March 24, 2020).

7   Results

7.1   Number of Party Supporters on VKontakte

To answer the first and third research questions we analyzed the statistics of traffic and communicative activity of communities of political parties and communities representing the selected interest groups on VKontakte. The main parameter we used to estimate audience size was the average number of views per post. In order to identify a clear artificial inflation of the number of views, we decided to measure the average number of likes, reposts, comments per post, as well as the integral indicator of audience activity – engagement rate (formula: (likes + comments)/views * 1000 [27]). The low value of this indicator will signal the artificial inflation of views compared to communities with approximately the same audience size, or those whose audience is smaller. Such a conclusion can only be made if the content of these groups has the same focus and subject matter (for example, conclusions based on a comparison of likes in entertainment and news communities will not be valid). Since we are comparing online communities of political parties, we believe that this condition is met.

Let us turn to quantitative characteristics of party communities in VKontakte, represented in Table 1. An indicator reflecting the audience size of communities in this statistic is the average number of views per post. According to this indicator, Navalny Team is in the lead, United Russia is in second place by a large margin, LDPR is in third, the CPRF in the fourth, Yabloko and Just Russia are in the last and second to last places. However, the engagement rate of posts in the communities of United Russia and the LDPR is significantly lower than in all other groups. In addition, the very close values of the average number of likes and reposts per post for United Russia (77 and 51, respectively) are atypical for posts in groups of political parties. This indicates an artificial increase in likes and reposts in this group.

If we assume that the views of posts in the communities of United Russia and the LDPR are cheated, then for a relative assessment of the popularity of online parties' pages on VKontakte, one can use an indicator that is difficult to artificially increase without anyone noticing. Of those available to us, such an indicator is the number of comments. As can be seen from Table 1, the leader in terms of the average number of comments per post is Navalny Team (which coincides with its leadership in the number of views per post); in the second place is the CPRF community. The rest of the parties occupy approximately the same level in this indicator.

To analyze interest groups, we created a classification and labeled each community. The detailed process of developing classification and labeling is presented in Appendix 2. We have identified three large categories of interest groups. The first category is various political associations. Their distinguishing feature is the presence of a declared ideology (for example, communist, nationalist, etc.). Both political movements and registered political parties fall into this category, except for the seven we are studying. The second category is protest communities. Users in these communities are united against decisions or policies of the authorities (for example, against the construction of landfills or the

Table 1. The size and activity of the audience of the Russian parties' communities on VKontakte

	Average views per post	Average number of likes per post	Average number of comments per post	Average number of shares per post	Engagement rate
United Russia	10,184	77	8	51	8
LDPR	7,361	64	10	9	10
Just Russia	2,345	91	6	8	41
CPRF	4,530	138	31	26	37
Yabloko	2,215	65	8	10	33
Navalny Team	34,264	767	105	53	25
Libertarian Party	2,480	55	11	4	26

demolition of architectural monuments). The third category is communities that defend their interests and rights. Their difference from the second category lies in the fact that their participants are not united against something specific, but seek their rights as a social group that is aware of its community. Examples of such communities are large families defending their benefits, trade union associations, defrauded equity holders, etc. For brevity, in what follows, these communities will be called "human rights".

The answer to the third research question is contained in Table 2, which shows the number of interest groups and the size of their audience. The first figure denotes the number of such groups found in the comments of the communities of political parties for the period under study, the second, in brackets, is the sum of the average views per post of these groups. According to these indicators, the CPRF (127 groups and a

Table 2. Interest groups in the comments of online communities of political parties on VKontakte

	Political associations	Protest communities	Human rights communities	Total
United Russia	2 (1,760)	1 (3,450)	2 (3,770)	5 (8,980)
LDPR	4 (1,527)	1 (3,450)	5 (11,834)	10 (16,811)
Just Russia	6 (4,292)	2 (4,436)	5 (13,012)	13 (21,740)
CPRF	78 (98,002)	22 (118,110)	24 (120,047)	124 (336,159)
Yabloko	3 (2,899)	4 (14,989)	5 (2,827)	12 (20,716)
Navalny Team	29 (76,006)	18 (52,128)	31 (107,816)	78 (236,000)
Libertarian Party	5 (9,091)	2 (7,782)	4 (7,340)	11 (24,213)

total audience – 336,159) takes the first place, the second – Navalny Team (75 groups and a total audience – 236,000). The number of interest groups in the comments of the remaining parties is in the range from 5 to 13, the total audience is from 8,980 to 24,213, which is significantly less than in the first two.

The quantitative distribution of interest groups revealed two centers of attraction for the public activity – the CPRF and Navalny Team. The number and general audience of interest groups found in the online communities of other parties was too small to be included in a comparative analysis. Therefore, in order to answer the second research question, we decided to analyze the composition of groups found in the communities of the CPRF and Navalny Team.

7.2   Interest Groups in the Communities of the Communist Party and Navalny Team

To answer the second research question about the qualitative characteristics of interest groups, we described and compared three types of the largest online communities, which we identified at the stage of community coding. This analysis was carried out on groups seeking support in the communities of two parties – the Communist Party (CPRF) and Navalny Team. We decided not to consider interest groups found in the comments of other online party communities, as their number turned out to be too small (from 5 to 13 groups). In addition, the total audience of these groups turned out to be ten times smaller than the audience of interest groups from the comments under the posts of the CPRF and Navalny Team.

Figure 1 shows the main interest groups – by the number of groups in the comments of party communities and by the size of the audience. Separately, we highlighted the groups that were present in the comments in both the CPRF and Navalny community.

Within political associations, there are two large ones, both in terms of the number of communities and the number of permanent audiences of the ideological blocs – communists and nationalists. Both blocks include various types of these ideologies. Communist communities are united by adherence to the ideas of communism or socialism in its (mythologized) Soviet form, as well as criticism of capitalism. Nationalist communities actively support the idea of a special role for the Russian people, which, as a rule, is accompanied by the endowment of Russians with special rights and restrictions on the rights of representatives of other nationalities and migrants.

Another type of political group that stands out against the background of others in terms of the number of communities is a variety of "left" movements. They are united by the idea of protecting and upholding the workers' rights. They call themselves social democratic and, unlike communist associations, do not postulate the Soviet Union as the desired model of society.

Figure 1 demonstrates a rather unexpected distribution of interest groups: despite the fact that the online platform of the CPRF attracts more communist associations, which corresponds to the very ideology of the CPRF, in terms of audience size, communist groups from the Navalny community are approximately equal to them (39,742 and 41,492 users, respectively). This picture results from the fact that in the comments under the posts of the Navalny Team appeared one communist union (38,913 users). At the same time, a larger number of nationalist associations with a big number of users who follow their activities are looking for supporters among the audience of the CPRF (6), and not Navalny Team (4). This fact can be interpreted in the sense that Navalny has become much less likely to be perceived as the supporter of the ideas of Russian nationalism by the nationalists themselves [40]. However, we should note that in three more nationalist groups, we could not count the audience, since during the period under study posts in these groups were not available.

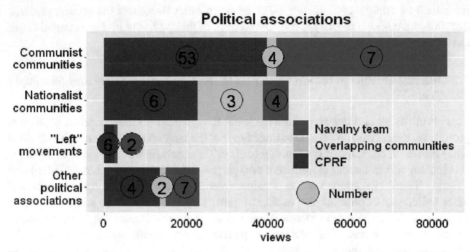

Fig. 1. Types of political associations from the comments in the communities of the CPRF and Navalny Team. The circles show the counts of political associations in each category.

Among the protest communities, the most numerous were the so-called "garbage" protests, as well as local urban protest communities (Fig. 2). The first ones are organized around the struggle of citizens with household waste dumps that pollute the environment. The latter are groups of citizens protesting against certain decisions and actions of city authorities, for example, the demolition of buildings. Moreover, in large cities, these online sites cover street protests against federal officials or the government.

Protest communities with fewer supporters turned out to be opponents of raising the retirement age (reform of 2018) and the current government (anti-Kremlin communities). The latter use the so-called "indefinite" protest as the main tool of resistance, the purpose of which is to remove the ruling political elite.

Figure 2 shows that the overwhelming majority of "garbage" protest groups, both in terms of the number and audience size, sought support in the CPRF community. The most numerous urban protest groups also appeared in the comments of the Communist Party community (4), although a larger number of such groups reached out to Navalny Team (9). The largest groups protesting against raising the retirement age sought support from both party communities. The majority, both in number (3) and in the audience size, of anti-Kremlin groups, communicated with the audience of Navalny Team.

Human rights associations appeared to be very diverse (Fig. 3). The most numerous in terms of the number of groups and the size of the audience were trade union associations. The parent groups, representing the interests of people with children, were the second

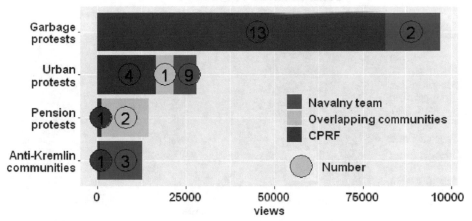

Fig. 2. Types of protest groups from comments in the Communities of the CPRF and Navalny Team. The circles show the counts of protest communities in each category.

largest. The third largest were groups supporting LGBT people and the feminist agenda, as well as a scparate category of groups representing judicial and legal associations. Among the judicial associations two types can be distinguished: those that provide legal support for various categories of people (for example, detained protesters), and those that unite people around a single trial and draw attention to it.

The majority of online trade union groups (Fig. 3) were found in the comments under the posts of Navalny Team (10). A large number of participants of trade union organizations that sought support from both Navalny's supporters and the CPRF is unevenly distributed across three communities: more than 80% of the participants belong to only one large trade union community (57,088 users).

The presence of numerous parental groups with a traditionalist bias (Fig. 3) in the comments of the CPRF (4), while under the posts of Navalny Team – the presence of communities supporting LGBT rights (2), allows us to draw a conclusion about the polarization of views on family relations and family policy among the supporters of these two political parties. At least this is how these largely opposing interest groups assess them.

A larger number of judicial and legal communities sought support among the audience of CPRF: 6 against 3 for Navalny. Nevertheless, in terms of audience size, these communities turned out to be approximately equal (Fig. 3).

In general, the set of interest groups found in the comments under posts in the CPRF and Navalny Team communities corresponds to the promoted ideology of these parties and the results of previous studies of their electoral potential, hence the fourth research question can be answered positively.

Communist and nationalist communities dominated among the political associations seeking support from the CPRF, which is in agreement with the ideas of the so-called "left conservatism", publicly broadcast by the leaders of the Communist Party [44]. The same groups were prevalent among the political associations appealing to the supporters

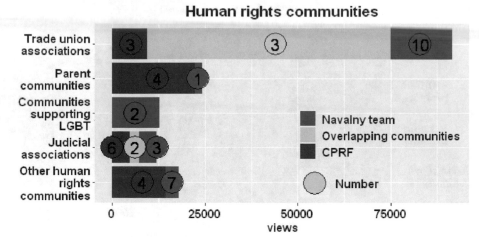

Fig. 3. Types of human rights groups from the comments in the Communities of the CPRF and Navalny Team. The circles show the counts of human rights communities in each category.

of Navalny Team, but their number was less than that of the CPRF (see Fig. 1). If the presence of nationalist associations can be explained by Navalny himself supporting a number of nationalists and nationalist ideas [10], the presence of communist associations can be interpreted as the desire of the opposition forces to find potential supporters from each other.

The distribution of protest groups between Navalny Team and the CPRF confirmed the fact that the latter still enjoys greater support in rural areas [29]. This is reflected in the fact that most of the groups protesting against garbage dumps near villages or small towns seek support mainly from the CPRF's online audience. Although the audience size in the urban protest groups appealing to the CPRF is close to that of the groups appealing to Navalny Team (see Fig. 2), the number of such communities is higher in Navalny Team's comments. In addition, the online community of Navalny Team also attracts a specific type of urban protest – the protest against the ruling elite (the so-called "indefinite" protest). This also aligns with the agenda promoted by the politician.

The discrepancy in the agendas of the two parties is also reflected in which human rights groups joined the discussion in the communities of the Navalny Team and the CPRF. Thus, most of the parental groups – opponents of juvenile justice and supporters of "traditional" views on raising children – sought support among the audience of the CPRF group, which corresponds to the declared traditionalism of this party [44]. While groups supporting LGBT people and the feminist agenda reached out to the audience of Navalny Team, which goes in line with Navalny's perception as a liberal politician [11].

The distribution of trade union associations between parties deserves special attention. The CPRF has been trying to postulate itself as a defender of workers' rights for a long time and, therefore, to attract trade union organizations to its side [33]. However, as our analysis shows, Navalny succeeded in gaining greater attention from unions (see Fig. 3). At the beginning of 2019, the politician announced the creation of his own trade union organization, ready to support various workers, but, first of all, state employees

[41]. The move likely prompted union organizations to communicate with Navalny's supporters, including online.

7.3   The Structure of Interest Groups' Online Communication

To answer the fifth research question, we used network analysis. We presented online communication of interest groups in party communities in the form of a graph (Fig. 4, the network was visualized in Gephi [2]), where nodes are parties and interest groups, and edges are links to interest groups in the comments of party communities (see the section Sampling). The size of nodes of interest groups reflects the actual audience (average number of views per post) of their online communities, the size of nodes of party communities corresponds to the total audience of interest groups that are associated with the corresponding party community. The weight (and, respectively, the thickness) of the edges corresponds to the number of links to interest groups that were found in the comments under the posts of the party community with which the edge links the interest group.

As can be seen from Fig. 4, most interest groups have a single edge, that is, each of these groups appeared in the comments of only one party. The most active community, which appears in the comments of several parties, is opponents of raising the retirement age (marked with the letter A on the graph). Although it is associated with all parties except the Libertarian, the graph shows that its communicative, and therefore political interest is focused on the audience of the Communist Party. Five more interest groups demonstrate the same behavior: they are associated with more than one party community, but the frequency of their appearance in the comments is greater either in the Communist Party communities or in the Navalny Team community, as can be seen from the thicker edge connecting the interest groups with the party. It is worth noting that there is also a noticeable number of interest groups that are more or less evenly present in the comments of various parties.

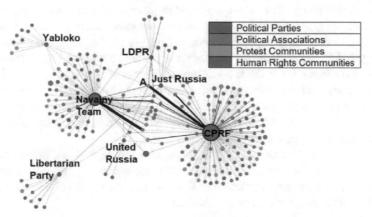

Fig. 4. Graph representing interest groups' links in the comments of political parties' communities

Nodes of interest groups with several edges may be viewed by parties as less loyal than those with only one party affiliation. However, they provide connectivity between the components of the graph, and consequently between different actors of the public discussion, preventing complete polarization and the formation of echo chambers. For parties, such communities are also valuable because, in a certain sense, they are their bridging capital [43]. They could attract supporters of other parties who are hesitant and change their party preferences.

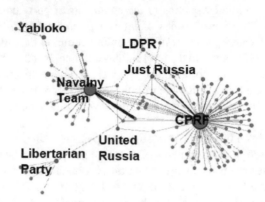

Fig. 5. Subgraph representing political associations' links in the comments of political parties' communities

To assess which types of interest groups make up the bonding capital to a greater extent, and which ones are bridging, we analyzed three subgraphs derived from the previous graph (Fig. 4). Each subgraph consists of nodes representing a party and one type of interest groups and edges between them (Figs. 5–7). If we visually compare these three subgraphs, we get the impression that political associations perform the cohesive function to a greater extent, since the respective subgraph is the only one of the three that has not lost the connectivity of all its components. However, this is a random effect related to the fact that political associations are the largest group. According to the average indicators of cohesion – density and average path length [14], the subgraph with protest communities is the most cohesive. Higher network cohesion is indicated by higher density and lower average path length. Thus, the respective indicators of the subgraph with protest associations are 0.045 and 2.843, of the subgraph with political associations – 0.02 and 2.855, of the subgraph with human rights organizations – 0.032 and 3.165 (the values were obtained in Gephi [2]).

Thus, the comparative analysis of subgraph cohesion showed that protest communities are more likely to establish bridging ties between parties, and political associations and human rights organizations are more likely to join only one political party, having a polarizing effect on political communication.

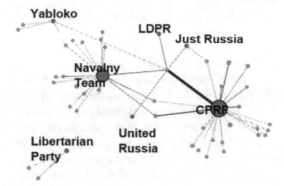

Fig. 6. Subgraph representing protest communities' links in the comments of political parties' communities

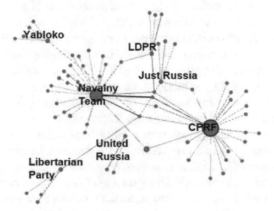

Fig. 7. Subgraph representing human rights communities' links in the comments of political parties' communities

8 Conclusion and Discussion

The study introduces a new method of using online communication in social media to assess the electoral potential of parties. We compared the electoral potential of the main Russian political parties using data from the online platforms of these parties on VKontakte from the end of March 2018 to May 2019. In this comparison, we used the typical quantitative indicators and our proposed indicator of interest groups.

In terms of the parties' communities' audience size, the most popular was the community of Navalny Team, a party representing the principal opposition and not having an official registration. The second and third places in popularity were taken by the online platforms of United Russia and the LDPR. However, part of their audience was apparently "artificial", since the activity parameters (engagement rate and average number of comments per post) in these communities were noticeably lower than those of other party communities, whose audience was significantly smaller. Therefore, we believe that the second place in this parameter should be taken by the Communist Party community.

The audiences of the remaining parties: Just Russia, Yabloko and the Libertarian Party of Russia, turned out to be approximately the same.

Our assessment that Navalny Team and the CPRF have the greatest electoral potential in social media is confirmed by another parameter we have proposed – interest groups that are looking for supporters in the communities of parties. In terms of the number of interest groups and the total size of their audiences, the Communist Party is significantly ahead of Navalny Team. In all other party communities, the number of interest groups turned out to be too low even to be included in the comparative analysis (see Table 2).

We have identified three types of interest groups: political associations, protest and human rights communities. Comparison of the aims and ideologies of these groups with the ideologies and historical background of the CPRF and Navalny's party showed almost complete correspondence. Among the audience of the CPRF, support is more often sought by conservative communities (opponents of juvenile justice), communist associations wishing to "return" to the USSR, and nationalists, which corresponds to the ideology of "left conservatism" [44]; as well as residents of poorly urbanized areas protesting against landfills near their settlements, who have traditionally supported this party. Navalny team, on the other hand, became a magnet for urban anti-Kremlin protests, LGBT supporters, and communist and nationalist communities and trade unions. The specificity of the communities similar to the CPRF can be explained by the fact that Navalny was pursuing a policy to attract these groups.

The high interest in Navalny Team on VKontakte can be explained by the fact that this party has been actively using social media for a long time to mobilize its supporters. The large number of interest groups with a fairly large online audience in the CPRF community's comments is explained, rather, by the presence of more opposition in the actions and statements of its leaders than in other parliamentary parties [34].

As the analysis of the communicative structure of online communities shows, some interest groups are included in communication with audiences of several party communities, and their activity may be unevenly distributed. That is, some interest groups are trying to attract supporters of different parties to their side. However, most often they turn to the audience of only one political force. This specific behavior of interest groups should be taken into account in assessing the electoral potential. Apart from this, it could be useful for parties to know which interest groups are bonding or bridging capital. In the studied case, protest communities more frequently performed a connectivity function and political associations and human rights communities – a polarizing one.

It should be recognized that the method for assessing the electoral potential of parties through public online communication of interest groups undoubtedly has limitations. Firstly, do not use SNS to mobilize their supporters, but use other mechanisms (television, pressure on groups of voters, non-admission of candidates) [20, 23], then they will remain out of our field of vision. Secondly, not all interest groups supporting a particular party will begin to communicate with the audiences of these parties in social media. For example, various business structures will prefer to avoid publicity. Thirdly, supporters of a particular party, its potential voters, are not always members of a particular interest group, but, for example, may simply identify themselves with the social group whose interests the party historically defends [21]. Fourthly, there are restrictions related to the availability of data from SNS.

Despite the shortcomings, our method of analyzing social media data allows us to use an additional source of information about the electoral potential of political parties. Moreover, due to the noticeable influence of interest groups on election results in both democracies and autocracies, our method for assessing their support for political parties in social media can be called universal for political regimes that have elections and public politics.

Acknowledgements. The study was conducted within the framework of the project No. 121062300141–5 "Comprehensive study of factors and mechanisms of political and socio-economic sustainability in the transition to a digital society".

Appendix

The first step in coding communities obtained from comments on the online party platforms was developing a classification. Since we did not have the opportunity to provide a reliable operationalization for the whole variety of groups that fall under the definition of "interest group", we used coding and categorization techniques to create a classification. Three characteristics became the benchmarks for the development of the initial set of categories:

1. Whether the group represents an organization or a person;
2. What is the main content of the group (political, news, entertainment, etc.);
3. The presence of goals that imply political participation.

Then we checked if there was an intersection or a hierarchy in the resulting codes. For example, we have combined all groups of schools, universities, colleges into one category – "educational institutions". As a result, we got a classification consisting of 71 categories.

In the second step, the groups were coded according to our classification. Since the formally defined features (name, brief description of the group) were not enough for coding the groups, it was also necessary to evaluate the agenda of the selected communities. The need to interpret the agenda forced us to use the qualitative-quantitative coding method (negotiated coding [8, 15]), in which the experts labelling the data had the opportunity to discuss their decisions. Two experts took part in the coding. They both labelled the complete dataset independently of each other. The accuracy was only 62%. The low accuracy appeared to be due both to the ambiguity in the interpretation of the agenda, and to the fact that a significant proportion of groups changed their subject matter or were blocked while the experts were conducting the coding. The groups whose codes did not match were re-coded by the experts through joint discussion.

As a result of the negotiated coding, we got not only more reliable group markup, but also improved the classification description. This allowed us to select from all categories exactly those groups that correspond to our definition of an interest group (see the section on interest groups and political parties). At the same time, we could not fail to note significant differences between the groups in terms of the type of interests defended.

References

1. Allcott, H., Gentzkow, M.: Social media and fake news in the 2016 election. J Econ Perspect **31**(2), 211–236 (2017)
2. Bastian, M., Heymann, S., Jacomy, M.: Gephi: an open source software for exploring and manipulating networks. In: Proceedings of the International AAAI Conference on Web and Social Media **3**(1), 361–362 (2009)
3. Baumgartner, F.R., Leech, B.L.: Basic Interests: The Importance of Groups in Politics and in Political Science. Princeton University Press, Princeton, NJ (1998)
4. Bawn, K., Cohen, M., Karol, D., Masket, S., Noel, H., Zaller, J.: A theory of political parties: groups, policy demands and nominations in American politics. Perspect. Polit. **10**(3), 571–597 (2012)
5. Baylor, C.A.: First to the party: the group origins of the partisan transformation on civil rights, 1940–1960. Stud Am Polit Devel **27**(2), 111–141 (2013)
6. Brand Analytics. Social Networking Sites in Russia: Figures and Trends, Autumn 2018. https://br-analytics.ru/blog/socseti-v-rossii-osen-2018/, last Accessed 24 Jun 2021
7. Cameron, M.P., Barrett, P., Stewardson, B.: Can social media predict election results? evidence from New Zealand. J Polit Mark **15**(4), 416–432 (2016)
8. Campbell, J.L., Quincy, C., Osserman, J., Pedersen, O.K.: Coding in-depth semistructured interviews: problems of unitization and intercoder reliability and agreement. Sociol Methods Res **42**(3), 294–320 (2013)
9. Central Election Commission of the Russian Federation. Election of Deputies of the Moscow City Duma of the Seventh Convocation. http://www.vybory.izbirkom.ru/region/izbirkom?action=show&vrn=27720002327736&region=77&prver=0&pronetvd=null, last Accessed 24 Jun 2021
10. Dollbaum, J.M., Semenov, A., Sirotkina, E.A.: Top-down movement with grass-roots effects? alexei navalny's electoral campaign. Soc. Mov. Stud. **17**(5), 618–625 (2018)
11. Dollbaum, J.M.: When life gives you lemons: alexei navalny's electoral campaign. Russ Anal Digest **210**, 6–9 (2017)
12. Durand, C., Blais, A., Vachon, S.: Accounting for biases in election surveys: the case of the 1998 quebec election. J Official Stat **18**(1), 25–44 (2002)
13. Eddington, S.M.: The communicative constitution of hate organizations online: a semantic network analysis of "Make America Great Again". Social Media + Society **4**(3), 2056305118790763 (2018)
14. Friedkin, N.E.: The development of structure in random networks: an analysis of the effects of increasing network density on five measures of structure. Soc Netw **3**(1), 41–52 (1981)
15. Garrison, D.R., Cleveland-Innes, M., Koole, M., Kappelman, J.: Revisiting methodological issues in transcript analysis: negotiated coding and reliability. Internet High Educ **9**(1), 1–8 (2006)
16. Gel'man, V.: Party politics in Russia: from competition to hierarchy. Europe-Asia Studies **60**(6), 913–930 (2008)
17. Gel'man, V.: Political opposition in Russia: a dying species? Post-Soviet Affairs **21**(3), 226–246 (2005)
18. Golosov, G.V.: Do spoilers make a difference? instrumental manipulation of political parties in an electoral authoritarian regime, the case of Russia. East Eur Polit **31**(2), 170–186 (2015)
19. Golosov, G.V.: The regional roots of electoral authoritarianism in Russia. Eur. Asia Stud. **63**(4), 623–639 (2011)
20. Golosov, G.V.: The september 2013 regional elections in Russia: the worst of both worlds. Reg Fed Stud **24**(2), 229–241 (2014)
21. Greene, S.: Social identity theory and party identification. Soc. Sci. Q. **85**(1), 136–153 (2004)

22. Grossmann, M., Dominguez, C.B.: Party coalitions and interest group networks. Am. Politics Res. **37**(5), 767–800 (2009)

23. Harvey, C.J.: Changes in the menu of manipulation: electoral fraud, ballot stuffing, and voter pressure in the 2011 Russian election. Elect. Stud. **41**, 105–117 (2016)

24. Hegelich, S., Janetzko, D.: Are Social Bots on Twitter Political Actors? Empirical Evidence From a Ukrainian Social Botnet. In: Tenth International AAAI Conference on Web and Social Media, pp. 579–582. AAAI Press, California USA (2016)

25. Hula, K.W.: Lobbying Together: Interest Group Coalitions in Legislative Politics. Georgetown University Press, Washington, DC (1999)

26. Jaidka, K., Ahmed, S., Skoric, M., Hilbert, M.: Predicting elections from social media: a three-country, three-method comparative study. Asian J. Commun. **29**(3), 252–273 (2019)

27. Judina, D., Platonov, K. Measuring agenda setting and public concern in Russian social media. In: Bodrunova, S.S. (eds.) International conference on internet science, pp. 211–225. Springer, Cham (2018)

28. Kennedy, C., et al.: An evaluation of the 2016 election polls in the United States. Public Opin. Q. **82**(1), 1–33 (2018)

29. Kolosov, V.A., Turovskii, R.F.: The electoral map of contemporary Russia: genesis, structure, and evolution. Russ Politics Law **35**(5), 6–27 (1997)

30. Kushin, M.J., Yamamoto, M., Dalisay, F.: Societal majority, facebook, and the spiral of silence in the 2016 US Presidential Election. Social Media + Society 5(2), 2056305119855139 (2019)

31. Libertarian Party. Libertarian Party: For Press. https://libertarian-party.ru/press, last Accessed 24 Jun 2021

32. Mamonov, M.V.: New risks of electoral sociology (Based on the Results of the Autumn 2013 Elections). Monitoring of Public Opinion: Econ Soc Changes J **1**(2014), 56–68 (2014)

33. March, L.: The contemporary Russian left after communism: into the dustbin of history? J Communist Stud Transit Politics **22**(4), 431–456 (2006)

34. March, L.: The Russian duma "Opposition": no drama out of crisis? East Eur Politics **28**(3), 241–255 (2012)

35. McCleskey, C., Nimmo, D.: Differences between potential, registered and actual voters: the houston metropolitan area in 1964. Soc. Sci. Q. **49**(1), 103–114 (1968)

36. Mediazona. May 5. Protest Action "He Is Not Our King". https://zona.media/online/2018/05/05/ne-tsar, last Accessed 24 Jun 2021

37. Mellon, J., Prosser, C.: Twitter and Facebook are not representative of the general population: political attitudes and demographics of british social media users. Res Politics **4**(3), 2053168017720008 (2017)

38. Mukhametshina, E., Korzhova, D.: Rally Against Blocking Telegram in Moscow Gathered Over 12,000 People (in Russian). Vedomosti. https://www.vedomosti.ru/politics/articles/2018/04/30/768321-miting-protiv-blokirovki-telegram, last Accessed 24 Jun 2021

39. Mukhametshina, E..: Rally Against the Isolation of Runet Attracted Over 15,000 People. Vedomosti (in Russian). https://www.vedomosti.ru/politics/articles/2019/03/10/795991-miting-protiv-runeta, last Accessed 24 Jun 2021

40. Myagkov, M., Shchekotin, E.V., Kashpur, V.V., Goiko, V.L., Baryshev, A.A.: Activity of non-parliamentary opposition communities in social networks in the context of the Russian 2016 parliamentary election. East Eur Politics **34**(4), 483–502 (2018)

41. Navalny. Our Very Big New Project: Starting a War for Your Salary (in Russian). https://navalny.com/p/6061/, last Accessed 24 Jun 2021

42. Navalny's Staff (in Russian). https://shtab.navalny.com/ last Accessed 24 Jun 2021

43. Putnam, R.D.: Bowling Alone: The Collapse and Revival of American Community. Simon and Schuster, New York (2000)

44. Rabotyazhev, N.: Between tradition and utopia: left-wing conservatism in Russia. Polis: Journal of Political Studies 4(4), 114–130 (2014)

45. Robertson, G.: Protesting putinism: the election protests of 2011–2012 in broader perspective. Problems of Post-Communism **60**(2), 11–23 (2013)
46. Rogov, K.: Political reaction in Russia and "party groups" in Russian society. Russ Politics Law **55**(2), 77–114 (2017)
47. Salisbury, R.H., Heinz, J.P., Laumann, E.O., Nelson, R.L.: Who works with whom? Interest group alliances and opposition. Am Political Sci Rev **81**(4), 1217–1234 (1987)
48. Tumasjan, A., Sprenger, T., Sandner, P., Welpe, I.: Predicting Elections with Twitter: What 140 Characters Reveal about Political Sentiment. In: Proceedings of the International AAAI Conference on Web and Social Media. AAAI Press, Palo Alto, California USA, pp. 178–185 (2010)
49. Turchenko, M., Golosov, G.: Smart enough to make a difference? An empirical test of the efficacy of strategic voting in Russia's authoritarian elections. Post-Soviet Affairs **37**(1), 65–79 (2021)
50. Van Deth, J.W.: A conceptual map of political participation. Acta Politica **49**(3), 349–367 (2014)
51. Vepsäläinen, T., Li, H., Suomi, R.: Facebook likes and public opinion: predicting the 2015 finnish parliamentary elections. Gov. Inf. Q. **34**(3), 524–532 (2017)
52. White, D.: Re-conceptualising Russian party politics. East Eur Politics **28**(3), 210–224 (2012)
53. Zhao, S., Shchekoturov, A.V., Shchekoturova, S.D.: Personal profile settings as cultural frames: Facebook versus Vkontakte. J Creative Commun **12**(3), 171–184 (2017)
54. Zhuravskaya, E., Petrova, M., Enikolopov, R.: Political effects of the Internet and social media. Annu Rev Econ **12**, 415–438 (2020)

Professional Networks

Societal Pressure or Free Choice: What Matters for Gender Composition of Informal Networks in the Workplace?

Deniza Alieva[1]([✉]) [iD], Sherzod Aktamov[2] [iD], Gulnoza Usmonova[3] [iD],
and Shukhrat Shadmanov[2] [iD]

[1] Management Development Institute of Singapore in Tashkent, Tashkent, Uzbekistan
dalieva@mdis.uz
[2] University of Digital Economics and Agrotechnologies, Tashkent, Uzbekistan
[3] University of Las Palmas de Gran Canaria, Las Palmas, Spain

Abstract. The following paper addresses the role societal norms (i.e., socially accepted behavior and gender-attributed stereotypes) play in shaping males' and females' social networks in higher educational institutions (HEIs). The universities were chosen as sources for network evaluations because of their proclaimed intention to promote gender-equality policies and a stereotype-free environment. However, culture and societal norms influence the representation of different genders in the workplace. The principal objective of the research is to evaluate how cultural characteristics and adherence to societal norms affect the gender composition of professional networks in formally gender-neutral environments in higher educational institutions. Namely, the paper attempts to determine whether the homophily of professional networks in HEIs depends on cultural characteristics. Independent variables include Masculinity vs. Femininity dimension (MAS) and Long-Term vs. Short-Term Orientation dimension (LTO) from Geert Hofstede's Cultural Dimensions model. Dependent variables include the E-I index of egos' networks. The data was collected in one Uzbek and one Russian university, with 52 participants in the first case and 49 in the second. The methods used consist of descriptive statistical analysis, the reconstruction of individual and aggregated personal networks, the measurement of six cultural dimensions from Hofstede's model, and the assessment of the level of homophily. The results of the study can help reveal problems men and women working in HEIs might face, specifically in terms of interacting within the institutions.

Keywords: Professional networks · Gender composition · Hofstede's model

1 Introduction

1.1 Interpersonal Contacts in the Workplace

The study of communication in everyday life and in the performance of labor activities reveals many situations when interpersonal contacts are of a forced nature and are carried out not only due to positive motivation, but also because of necessity and obligation.

Many connections are established and maintained in organizations. The success or failure of an individual within an organization and its future depend on many factors. One factor is the understanding of communication flows in an organization that can provide more information on the makeup of a company.

An organization is first created, then filled with staff who begin to build interpersonal relationships with one another. The degree of convergence of members' personality traits in the organization with its program, structure, culture, and their job functions affect the organizational performance and goal achievement [2, 4, 65]. In addition, relationships in the company and their characteristics, i.e., gender composition of networks [5, 11, 42, 46, 50], shared cultural values [29, 59], knowledge [48, 61], and educational background [44, 50], etc., might act as direct influencers or moderators of the performance of employees. The exploration of such connection and proofs of its importance are leading companies to adopt different policies, including gender-equality ones. Better understanding of the structure of gender composition in different organizational and cultural environments might increase policies' effectiveness and improve their implementation in organizations.

Considering this, the present research aims to understand the structure of interpersonal relations in formally gender-neutral environments in higher educational institutions (HEIs). It attempts to determine the gender homophily of networks and its potential connection with cultural characteristics of organization's members.

1.2 Formal and Informal Networks in Organizations

Social network analysis focuses on two types of networks in the workplace. First are formal networks – ones that are established by hierarchy in the organization. They are regulated by the internal structure of the company and they formalize relationships within the organization [9]. They are based on the organigram of the company with its functional departments, communication channels, policies, and procedures. They are relatively stable and long-lasting [67].

The second type of network in focus are informal ones. They emerge from interaction between co-workers that happen while they are dealing with work tasks [52]. Informal networks complement and add extra value to formal structures, without being explicit and visible [9]. It is agreed that the work process coordination, communication flows, and other non-formalized activities often take place through informal relationships [18]. Informal relationships are important in decision-making processes [60, 68]. They have influence on the performance of employees [39, 66], their perception of the conditions of work [47], career satisfaction [73], and attitude towards self-development initiatives promoted by the workplace [55, 72]. Moreover, their impact is noticed not only at the individual level, but also in the whole organizational network, as its members have influence on each other [22] and distribute power between one another [3].

At the formal level, members of the organization act based on the goals of the organization. At the informal level, they act because of their personal interests or the interests of informal groups. At the same time, formal goals and informal interests do not always coincide [28].

Informal relationships may be more significant for members of an organization (or a separate unit) than formal, officially prescribed ones; moreover, it is impossible to influence the creation of informal structures or prevent their occurrence. At the same time, in many cases, informal relations can be used to facilitate the solution of formally set tasks.

Our research concentrates on informal networks as opposed to formal networks due to their ramifications for informal workplace division. However, the difference across formal and informal networks is one of extent rather than form. Formal networks are often public, authorized, and have distinct limits, while informal networks are typically private, voluntary, and have porous bounds. Formal work networks, for example, tend to be formally accepted by companies and concentrate on reaching a labor or corporate objective. Additionally, they often have recognizable affiliation and a stated organization. In comparison, participation in informal networks is neither formally controlled nor publicly acknowledged. The purpose of such networks may be professional, private, or social [23]. The contrast between formal and informal networks is crucial to this investigation since exclusion from formal networks has varied consequences from exclusion from informal networks. A worker who is barred from a formal network might claim that she or he has been mistreated by citing corporate policy or written job specifications. In fact, employees who are barred from informal networks have limited opportunities for exposing mistreatment since, in general, employers do not assume responsibility for informal work relationships.

Although scholars have studied this phenomenon at length, the informal organization's culture, the "shadow structure", in the workplace remains obscure [1].

1.3   Gender in Informal Networks in Organizations

While theorists have postulated that gender is ingrained in work organizations, empirical evaluations often focus on one side of organizational life – the formal one [40]. Over 20 years of studies have demonstrated that the gender and race of employees impact their official status and compensation in the workplace. For example, women and people of color have positions with lower status, authority, promotion opportunities, and compensation than males and Whites [15, 20].

According to the status construction hypothesis [17], while building networks, employees do take each other's gender, race, and ethnicity into consideration. In particular, people rank one another according to the resources they control; the more resources a person has, the more capable others perceive him or her to be. Workers may believe that a person's gender, color, or ethnicity indicate their value and abilities since these factors are tied to an individual's resources [69].

Additionally, network scientists acknowledge that employees' preferences influence with whom they develop network links [16]. However, these models stress different causative variables influencing network composition. According to network researchers, comprehending network structures hinges on structural elements. Consequently, they would assume that the impact of employee gender on the status of network members is indirect and mediated by structural variables. In contrast, status construction theorists would suggest that workers' ideas on gender are crucial network composition drivers.

Therefore, the gender of employees should have direct or interactive consequences on the status of network participants.

Moreover, we have noticed that gender might lead to exclusion in different types of networks in the workplace. The examples include such sectors as construction [14], information and communication technologies [30], textile and clothing [45], and even academia, with focus on organizations in general [26, 51, 63], or particular groups of people [24, 71]. These studies admit that informal networks contribute to the maintenance of inequality, but few study the particular processes through which the gender of workers influences their informal networks [52]. Therefore, our knowledge of informal differentiation in workplace settings falls behind that of formal differentiation [43].

Further study is required to comprehend how gender is integrated into the informal structure of workplace groups [13, 19]. The importance of informal networks in employment disparity might be better understood if studies on network creation and transformation were conducted. Understanding how individuals build network links and how these networks evolve over the course of their careers might help researchers pinpoint the times at which network exclusion is most likely to occur and the influence it has on the following job outcomes of workers.

1.4   Cultural Dimensions in Geert Hofstede's Model

The term "culture" is studied by many scientific disciplines, being differently explained by various authors. Culture is a group phenomenon – it is shared by people that are present in the same social environment and interact with each other on a regular basis [36]. According to Eliot [27], the birth of culture leads to the appearance of a hierarchy. If applied at an organizational level, it determines power relations and assists to establish power balance [34, 56].

The studies analyzing direct relationship between gender diversity and organizational culture [25, 31] or the mediating role of cultural features in connection between gender and organizational performance [58, 64] confirm the importance of organizational culture for understanding of gender diversity in the workplace.

Geert Hofstede created a model of six cultural dimensions that are widely used in the studies of organizational culture. In his model, Hofstede [35] shares the idea that cultures have common problems and therefore common cultural dimensions. Based on these assumptions and the findings obtained following his research, he determined six universal categories, or dimensions, whose degree of expression can serve as a differential culturological feature of organizations: Power Distance, Individualism vs. Collectivism, Masculinity vs. Femininity, Uncertainty Avoidance, Long-term vs. Short-term Orientation, and Indulgence vs. Restraint.

It is worth stating that since its founding, the Cultural Dimensions model [35] has served as a reference in a large number of investigations in the context of international business [49] and other organizational studies, including those that evaluate the impact of culture on formal and informal organizational structures [41, 62].

Based on the definition of the model's author, the following brief descriptions of the dimensions might be provided [38].

The first dimension – Power Distance – measures the distance in relation to power. It is expressed through the attitude to inequality, uneven distribution of power and social benefits within society and particular social institutions. The second dimension – Individualism vs. collectivism – indicates the degree of integration of the individual into the group, allowing to compare more individualistic and more collectivistic cultures. The third dimension – Masculinity vs. Femininity – reveals the degree of differentiation of the role behavior of males and females. The fourth – Uncertainty Avoidance – reveals the degree of tolerance of representative cultures of interest in relation to ambiguity, and the degree of their willingness to take risks. The fifth dimension – Long-term vs. Short-term orientation – was revealed after the analysis of Asian cultures. In cultures characterized by its extreme form – Long-term orientation – diligence, perseverance, and persistence are considered to be the virtues. They are seen as virtues that can help create a better a prosperous future. However, the cultures characterized by Short-term orientation focus on present and past. Social life is characterized by relationships determined by status, tradition, and the desire to preserve the established order. Finally, the sixth dimension – Indulgence vs. Restraint – evaluates the readiness of society to enjoy life and have fun.

For the purposes of the current research, we will focus on two of the six dimensions – Masculinity vs. Femininity (MAS), and Long-term vs. Short-term orientation (LTO). These were selected as they potentially might be connected with gender homogeneity of respondents' networks. In high masculine cultures, gender role differentiation becomes clearer as competition and rivalry is encouraged. Moreover, there is an orientation towards "male" values: perseverance and self-confidence. In feminine cultures, modesty, restraint, sympathy for the weak, and social orientations toward activities are valued. The effect of MAS on structures and characteristics of relationships was measured and evaluated in several studies, proving the existing connection. The researchers focused primarily on the analysis of social capital [8], homogeneity of users in social networking sites [33], and its connections with appointment practices in Human Resources Management [70]. Meanwhile, studies on the effects of LTO on homogeneity often look at consumer behavior [21], corporate behavior [7], and profit changes [32].

There is a limited number of studies that determine existing connection between importance of traditions in the workplace, that forms part of LTO cultural dimension, and gender diversity [6, 53, 57]. In their works the researchers show that preservation of traditions and focus on them as bases for the prosperous future might be connected with lack of gender diversity in organizations. In contrast, gender equality is observed in those companies where the emphasis is put on adaptation of new traditions and evolution of the old ones, as it is done in long-term oriented cultures.

Based on this research and the results of these studies, we can expect that high MAS leads to higher gender homogeneity, while high LTO is connected with higher levels of heterogeneity. Understanding if the same pattern can be observed in the circumstances of formally gender-neutral HEIs may cover the research gap in the field, as there is little research on it to date.

1.5 Objectives and Scheme of the Study

The present study aims to evaluate the influence of cultural traits on the circle of contacts employees of HEIs have. The *principal objective* is to evaluate how cultural characteristics and adherence to societal norms affect the gender composition of professional networks in a formally gender-neutral environment. Based on this, the following *specific objectives* have been suggested:

- To measure cultural dimensions in two HEIs;
- To reconstruct professional networks of academic and administrative staff of these HEIs;
- To evaluate the influence that cultural dimensions have on gender composition of the professional networks.

Using Hofstede's model dimensions, we aim to analyze the effect culture might possibly have on personal networks' homophily levels in terms of gender. As mentioned before, two of the dimensions were chosen for the purposes of the current research. First, it is the Masculinity vs. Femininity (MAS), as it is related with inflexibility of gender-based roles in the society. Second, the Long-term orientation (LTO), that measures the importance of traditions and openness for changes among people, including the shifts in gender-based roles and occupations. Based on literature review and our field observations, the following hypotheses are formulated:

Hypothesis H1: The higher the MAS index among employees, the more gender-homogeneous the personal network in the workplace.

Hypothesis H2: The higher the LTO index among employees, the more gender-heterogeneous the personal network in the workplace.

2 Methodology

2.1 Questionnaire

The study used an online self-administrated questionnaire that participants had to fill out. Russian was used as the only language of data collection, as all respondents reported speaking it before receiving the survey. The questions were divided into several groups – each of them collected data aimed to fulfill the objectives.

The first part was focused on several characteristics of the participants. In particular, gender, work experience in HEIs, experience living or studying abroad, and type of professional activity (academic or administrative) were measured. The second part contained the Values Survey Module (VSM 2013) [37], that collected information on six cultural dimensions of participants. Finally, the third asked participants to provide a list of 20 colleagues with whom they maintain contact in their HEI. The number of *alteri* was limited to 20 because of the feedback provided by participants. For the majority of those who received the questionnaire, it was impossible to name more people with the characteristics required. However, all participants were able to name 20 *alteri* without any pressure or suggestions.

For each *alter* the information on his/her connection with others was collected from the ego. Specifically, they had to mark if *alteri* know each other, following the scale where 0 stands for "do not know each other", 1 – "know each other", and 2 – "are friends". In addition, the participants provided the following information: gender, department (same as ego or not the same), closeness to ego, and type of work (academic staff or administration).

At the end of the questionnaire, respondents could agree to share additional information in form of a critical incident relevant to the study. If willing, we contacted them and conducted semi-structured interviews to obtain further details. Such approach helped to enrich the data collected and to analyze particular experiences of participants.

2.2 Data Collection and Data Analysis

Data collection was conducted from 2020 to 2021. First, we contacted several Uzbek and Russian HEIs in order to evaluate their interest in participating in the study. Then, two universities that had characteristics relatively similar to each other were found. The total number of employees (academic staff and administrative staff included) in Uzbek HEIs is 429 people, in Russian – 447. Another variable controlled – gender composition of academic and administrative staff – was almost equal for both cases. 50.8% of employees in Uzbek HEIs are males, while in case of Russian HEIs, that number reaches 49.0%.

We sent an email to each prospective participant explaining the objectives of the study and asking for confirmation of willingness to participate. Following the confirmation, the questionnaires were shared with respondents, and each participant had two weeks to provide answers. After this period, the first reminder was sent, giving the participant two additional weeks to fill out the questionnaire. Finally, the second reminder was provided, giving another week to complete the survey. Such approach significantly reduced the number of non-answered questionnaires, and the overall response rate reached 84.2%.

Participants willing to share critical incidents were contacted two weeks after the first stage of data collection finished. We had Zoom interviews with them and transcribed the conversations afterward.

The data analysis was conducted with IBM SPSS Statistics 23 and Ucinet 6 [10]. The visualizations of networks were done by utilizing visone-2.23 [12]. We calculated the descriptive statistics and general network characteristics for each professional network. The cultural dimensions were calculated using Values Survey Module 2013 Manual [38]. The following formulae were used for each of the dimensions, where m_{XX} stands for the score for question XX in the survey. Each index has a range of about 100 points, and the C(**) is a constant that does not affect the comparison between countries [38].

Power Distance Index (PDI):

$$PDI = 35(m_{07}-m_{02}) + 25(m_{20}-m_{23}) + C(pd) \qquad (1)$$

Individualism Index (IDV):

$$IDV = 35(m_{04}-m_{01}) + 35(m_{09}-m_{06}) + C(ic) \qquad (2)$$

Masculinity Index (MAS):

$$MAS = 35(m_{05}-m_{03}) + 35(m_{08}-m_{10}) + C(mf) \qquad (3)$$

Uncertainty Avoidance Index (UAI):

$$UAI = 40(m_{18} - m_{15}) + 25(m_{21}-m_{24}) + C(ua) \qquad (4)$$

Long-Term Orientation Index (LTO):

$$LTO = 40(m_{13}-m_{14}) + 25(m_{19}-m_{22}) + C(ls) \tag{5}$$

Indulgence vs. Restraint Index (IVR):

$$IVR = 35(m_{12}-m_{11}) + 40(m_{17}-m_{16}) + C(ir) \tag{6}$$

The level of homophily of the networks was measured by using the E-I index [47], which can range from −1 to 1. The index was calculated using Ucinet software [10], where for each network the subtraction of the number of in-group ties from the number of out-group ties was performed, divided by the total number of ties. A −1 indicates complete homophily, or when an individual has relationships with only actors of the same characteristics that he/she has. An E-I score of 1 stands for complete heterophily, when all the *alteri* possess different characteristics in comparison with the ego. An E-I score of 0 means that an equal number of *alteri* share and do not share the characteristics of the ego.

Finally, the regression model between the two cultural dimensions and the E-I index of the personal networks for each country was reconstructed to evaluate the influence of LTO and MAS on the level of homophily of participants' personal networks.

2.3 Sample

As a study sample, we used two HEIs: 52 people working at a university in Uzbekistan, 49 in another university in Russia. In each group the distribution by gender and occupation are statistically equal. From the first group of participants, 27 (51.92%) work as lecturers, professors or instructors, and 25 as administrative staff (48.07%). The difference is statistically not significant with Chi-square 0.076, p-value 0.99. In the second group, 24 participants (48.98%) are academic staff, and 25 (51.02%) administrative staff of HEIs (the difference is statistically not significant with Chi-square 0.0001, p-value 0.99). Due to the reduced number of participants, no grouping by position or rank was made.

48.51% of the total sample are male, and 51.49% female. The average number of years they work in the current HEI is 7.8 (SD = 2.4), with Russian participants more long-lasting in their place of work (M = 10.4, SD = 1.6).

3 Results

3.1 Cultural Dimensions

The results of the analysis of VSM 2013 are presented in Table 1. In general, in all dimensions there is a statistically significant difference between the Russian and Uzbek samples. Based on these results, we can make a number of conclusions about the common traits of participants from both Russian and Uzbek samples: (1) they perceive power distance in their workplace, (2) prefer to avoid uncertainty and plan ahead with a high degree of certainty, (3) do not allow themselves considerable leisure time, and (4) limit consumption and the spending of money overall.

Table 1. Cultural dimensions

		PDI	IDV	MAS	UAI	LTO	IVR
Mean	Russia	92	43	33	90	83	25
	Uzbekistan	95	28	83	94	27	27
SD	Russia	3.27	4.01	3.69	1.21	2.80	1.20
	Uzbekistan	2.90	2.62	3.58	2.56	2.33	2.77
	t-test	−4.86	22.09	−69.03	−10.10	108.64	−4.74
	p-value	0.00**	0.00**	0.00**	0.00**	0.00**	0.00**
	df	95.93	81.95	98.20	73.68	93.64	70.33

Note: PDI – Power Distance, IDV – Individualism, MAS – Masculinity, UAI – Uncertainty avoidance, LTO – Long-term orientation, IVR – Indulgence vs. Restraint
**significant at 5%

There is a slight difference (43 versus 28) in Individualism. It confirms that the Uzbek sample have a more collectivistic approach in the workplace, perceiving the importance of the group and its opinion more than the Russian sample.

The difference between indices gets more notable in two remaining dimensions. Uzbekistan scores higher in Masculinity (33 vs. 83), showing that the respondents from that country have more established and distinct masculine and feminine roles. In such cases, men tend to focus on material success and be assertive and tough, while women are supposed to be tender and modest. The focus of females should be the quality of life from a spiritual and immaterial perspective, and males should provide the material base for it.

The difference in Long-term orientation (83 vs. 27) points out that among Russian respondents, the orientation toward future rewards and perspectives is more relatable. These participants often plan more ahead and do not look back to the past as much, focusing more on upcoming challenges and goals. In the case of Uzbekistan, participants demonstrate an interest in preservation of traditions and fulfilment of social obligations, which is higher than that toward goals established for the future. The desire to "save face" in society can be the reason for following intentionally or unintentionally the traditions (including those of gender roles).

3.2 Professional Networks

The reconstruction of professional networks lets us understand their composition and evaluate their structural characteristics. After examining participant networks, we found that in both Uzbek and Russian HEIs, members include faculty and administrative staff in their personal networks. In the case of the Uzbek HEI, networks are more homogenous. Of the total *alteri* of 52 networks, 65% (n = 676) are representative of the same professional occupation as the ego. Among Russian university representatives, such value is 45% (n = 441).

The density of all networks is 1 and there is no fragmentation in any network as the egos were asked to name the *alteri* from their closest professional circles. Subsequently, all *alteri* know each other, with the average distance being very low (M = 1.00, SD = 0.00), and the Average Degree of 19 for all the networks.

From the personal networks, two-mode networks of *alteri* gender divided by professional occupation were built. After that, the aggregated networks were created in order to see the overall composition of males' and females' professional networks in Uzbek (Fig. 1) and Russian HEIs (Fig. 2).

Among Uzbekistan participants, 27 are females and 25 males. The female aggregated network demonstrates the prevalence of female *alteri* with 65% of the total (n = 351). The same-gender composition is observed in male aggregated networks; however, the percentage is lower – 56% (n = 280). The average E-I index for males' personal networks is –0.271, for females it is –0.347.

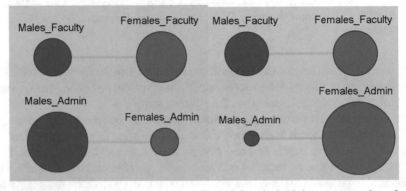

Fig. 1. Clustered graphs of Uzbek males' (left) and females' (right) aggregated professional networks, where the color of the node represents the gender of *alteri* (blue – male, pink – female), the size of the node – number of *alteri* with same characteristics.

If analyzed not only by gender, but also by professional occupation, the aggregated networks demonstrate different perspectives, especially in the case of male egos. Among Uzbek male academic staff, the aggregated gender distribution reaches the proportion of 42.5% males (n = 102) and 57.5% females (n = 138). The average E-I index of male faculty personal networks is 0.15. Male administrative staff increases the presence of male *alteri* in their professional networks: 68.46% males (n = 178) and 31.54% females (n = 82), with the average E-I index at –0.369. Female academic staff maintain lower percentage of female *alteri* – 50.67% (n = 152) with an average E-I index of – 0.013. The presence of females in professional networks is more remarkable in the case of administrative staff – 82.92% (n = 199), with the average E-I index at 0.658.

Russian participants demonstrate more equal distribution in terms of gender in their personal networks based on professional relationships (Fig. 2). In particular, among male respondents' *alteri* 53.33% (n = 256) are males and 46.67% (n = 224) females. The average E-I index for male personal networks is 0.008, for females 0.028. The analysis by gender combined with professional occupation does not show remarkable changes in distribution. Male faculty counts for 48.18% (n = 106) of males in their professional networks, and male administrative staff 57.69% (n = 150). Female faculty has 52.69%

(n = 137) of female representation in their networks, while female administrative staff counts for 55.83% (n = 134).

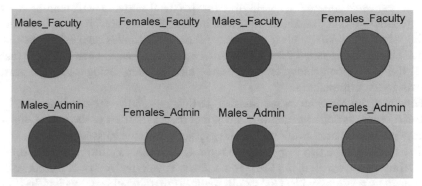

Fig. 2. Clustered graphs of Russian males' (left) and females' (right) aggregated professional networks, where the color of the node represents the gender of *alteri* (blue – male, pink – female), the size of the node – number of *alteri* with same characteristics.

3.3   Influence of Cultural Dimensions on Professional Networks

After conducting analysis of the personal networks' composition, the study on cultural dimensions' influence was completed. To do so, we chose two dimensions that apart of being connected to traditional or progressive approach to gender in society, had shown relatively bigger difference between countries, namely Masculinity (MAS) and Long-term orientation (LTO). Then, the computation of regression between the E-I index and the selected cultural dimensions was performed.

In order to check the relationship between LTO and MAS and E-I index, we ran a regression model. In this model, the E-I index serves as a dependent variable, which is a proxy for the homophily of the personal networks of participants. Our independent variables are LTO and MAS, measured using Values Survey Module 2013 Manual [38].

We constructed two models: one for Uzbekistan and one for Russia. The model for Uzbekistan shows that the model is solid enough based on Gauss-Markov criteria. The independent variables (LTO index and MAS index) have a statistically significant impact on the E-I index (the significance level is within 1% α). Also, the coefficient of determination is equal to 0.76, which indicates that these two variables are good predictors of the E-I index. Besides, the F-stats p-value also shows that these two variables combined have statistically significant impact on the dependent variable.

In the second model, where the sample data was collected in the Russian HEI, the coefficient of determination is 0.85. The F-stats p-value is significant (0.28), and the t-stats p-value also shows the statistical significance. The model for Russia also shows that the model is solid enough based on Gauss-Markov criteria. We accept this model as reliable having MAS and LTO indices as good predictors of the networks' level of homophily (significance level is within 5% α).

Hypothesis H_1: *The higher is MAS index among employees, the more gender-homogeneous the personal network in the workplace.*

The MAS index has a negative impact on the E-I index. This means that, in Uzbekistan, if the ego's masculinity is higher, the personal network will tend to be more homogeneous in shared characteristics with the ego. For example, if the ego is male, he, scoring high in the MAS index, might have more males in his network in the organization. The H_1 is supported for the case of Uzbekistan. Thus, we can conclude that the higher the MAS index among employees, the more gender-homogeneous the personal network in the workplace is.

However, the MAS index for Russian participants shows the opposite result. The regression analysis demonstrates that the higher the MAS index of the person, the more heterogeneous his/her network will be. H_1 is rejected for the Russian HEI.

The results of the two samples give room to think that culture plays the role of a moderating factor which is impacting on a decision who the actors of the network are forming their relations with. Our results indicate that Uzbekistan society has higher masculinity index compared to what is in Russia. We can assume that there is a limited choice for a person to form his/her network depending on gender if the society has certain established expectations. A person who scores high in MAS and who in his surrounding has the same high-masculinity-type of people tends to behave according to the rules of masculine cultures. In forming the networks, the factor that the culture itself is masculine and requires certain patterns of behavior inhibits one to establish more heterogeneous relations.

In a masculine culture, during socialization boys and men are encouraged to reject or avoid anything stereotypically feminine. The threat of being judged by society and "poor" contact from women arising from stereotypes about appropriate woman's behavior force men to construct their network in a more homogeneous way. Such a network may allow to accomplish societal expectations of men who can be tough and aggressive, suppress emotions (other than anger), distance themselves emotionally and physically from other men, and strive toward competition, success and power. Our results coincide with the findings of Mehta & Strough [54] where actors in the network may mutually benefit as their similarity helps to reduce the costs of communication, provides them opportunities to join similar activities.

Hypothesis H_2: *The higher the LTO index among employees, the more gender-heterogeneous the personal network in the workplace.*

For Uzbekistan, the effect of the LTO index on the level of homophily is positive. It shows that if the Uzbek ego scores low, the personal network in the workplace can be expected to be more homogeneous. The H_2 is supported for the Uzbek HEI.

The results of the second model with the Russian sample in terms of the LTO index's impact on the E-I index does coincide with the results from Uzbekistan's sample. In the case of a participant from the Russian HEI with high LTO index, for example, a highly heterogeneous network can be expected. H_2 is supported in the case of Russia.

Thus, we can conclude that the long-term orientation of an individual makes networks more gender-heterogeneous, while short-term orientation forces to construct the network with a homogeneous nature. In short-term oriented cultures, where achievements of today are more valuable than future rewards, a homogeneous network in the short-term may help to reach individual's goals. We can't claim that there is a causality between these two factors. It is still a question if the short-term orientation causes the networks to be more homogeneous or if homogenous networks are the ones that impact the individuals to behave in a short-term oriented manner due to the high pressure of surroundings. In such cultures as in our second sample (Russia) where orientation is given towards future investments and value is given to persistence, perseverance, and adaptability there are more chances to achieve goals when an individual can enjoy a higher variety of his/her network, even if it is only in terms of gender.

The results of hypotheses testing demonstrate that the effect of gender equality policies, applied in the HEIs in question, is minor. The national culture of participants structures their relationships in the workplace, contrasting with expected effects of adopted policies.

4   Limitations and Suggestions for Future Research

While we used solid theory taken from relevant literature, the conducted research has limitations worth recognizing. The study used a limited number of participants; however, it can be seen as representative of one local university in capitals of Russia and Uzbekistan. To expand the understanding of the phenomenon discovered, additional data may be collected in other local or international universities both in the capitals and in the regions. Being a large heterogeneous country, Russia will require more profound analysis, as in many territories within the country where different cultures may be observed. Nevertheless, the current research accomplished its objectives and has improved the understanding of connections between culture and personal networks structures.

We focused on two of six dimensions of Geert Hofstede's Cultural Dimensions model. It would be interesting to explore the effect of the other four dimensions or the combinations that several of them may have on gender homophily of social networks in HEIs. Moreover, for the purposes of the research, only information about binary gender was collected. This decision to avoid the mention of non-binary genders was taken due to societal and cultural norms in both countries. Such omission helped to increase the comfort level of the participants during self-administered data collection and to attract more people to the research. However, the addition of a non-binary gender category could change the focus of the study.

The rejection of H_1 in the Russian HEI case can be explained by factors other than cultural dimensions that potentially might affect the composition of networks, such as the educational background of participants and experience in living abroad. The detection of discrepancy between the two HEIs invites further research for more profound analysis and determination of these additional factors.

The methods applied in the study also have limitations to be addressed in the future. First, the information regarding connections between *alteri* was collected from egos, which potentially can contain errors. The egos might have an incomplete knowledge of *alteri*'s networks, which could affect the data. We do not expect the error in networks recreation to be frequent, as the respondents included in their networks 20 people closest to them from their workplace, thus reducing the margin of error.

Second, we did not analyze the combination of the effect of cultural dimensions with other factors that may influence gender composition of networks. Only the total number of employees in HEIs and their gender composition were controlled. However, the exploration of other factors (such as educational background of *alteri,* their experience abroad, their perception of gender equality, etc.) and their effects might lead to more research in the field, expanding the body of knowledge in the area.

5 Conclusion

The results of the study contribute to the existing literature by demonstrating that there is a connection between at least two dimensions of Geert Hofstede's Cultural Dimensions model and gender homogeneity in HEIs. They prove that the cultural values and the societal norms play a certain role in the formation of personal networks in HEIs in Russia and Uzbekistan. In particular, the clear determination of gender roles and the traditional approach to their definitions, combined with the desire to follow tradition, leads to structural changes in the case of Uzbek HEIs. The networks become more homogeneous with the increase in the importance of traditional values and more rigid definitions of gender roles. In Russian HEIs, the fostering of virtues oriented towards future rewards is connected with higher levels of gender diversity in personal networks. However, less defined gender roles increase the chances to have more heterogeneous personal networks.

Such conclusions demonstrate that although in the cases of Uzbekistan and Russia, gender-equality policies and the involvement of both gender representatives are used widely in higher educational institutions, the effect of culture is still present on a large scale. It can create a cultural shock for people from other cultures, where this effect is not observed. In addition, such situation can somehow disrupt the communication flows in HEIs and influence their quality and consistency. Moreover, those who do not demonstrate such patterns of behavior as determined by the present study, might experience problems with fitting in.

In contrast, the gender homogeneity might be positive for representatives of the cultures in question or ones with similar characteristics. Being used to have such pattern of behavior, they might find it easier to adapt to the environment of HEIs. In addition, such structure of relationship might potentially be a source of support and help in ease of communication between people of same gender occupying positions of different ranks in HEIs.

References

1. Al-Twal, A., Cook, C.: Explaining the power of informal networks in academia: a Jordanian perspective. J. Furth. High. Educ. **46**(6), 807–821 (2022)
2. Arifin, A.H., Saputra, J., Puteh, A., Qamarius, I.: The role of organizational culture in the relationship of personality and organization commitment on employee performance. Int J Innovation Creativity Change **9**(3), 105–129 (2019)
3. Atkinson, S. R., Moffat, J.: The agile organization: From informal networks to complex effects and agility. Assistant Secretary of Defense (C3I/Command Control Research Program), Washington D.C. (2005)

4. Awadh, A.M., Wan Ismail, W.: The impact of personality traits and employee work-related attitudes on employee performance with the moderating effect of organizational culture: the case of Saudi Arabia. Asian J Bus Manage Sci **1**(10), 108–127 (2012)
5. Bae, K.B., Skaggs, S.: The impact of gender diversity on performance: The moderating role of industry, alliance network, and family-friendly policies–Evidence from Korea. J. Manag. Organ. **25**(6), 896–913 (2019)
6. Baines, S., Wheelock, J.: Work and employment in small businesses: perpetuating and challenging gender traditions. Gend. Work. Organ. **7**(1), 45–56 (2000)
7. Baker, W.E.: Market networks and corporate behavior. Am. J. Sociol. **96**(3), 589–625 (1990)
8. Batjargal, B.: Comparative social capital: networks of entrepreneurs and venture capitalists in China and Russia. Manag. Organ. Rev. **3**(3), 397–419 (2007)
9. Benatti, L.N., dos Santos, A.F., Trindade, E.M.D., Gomes, N.L., Prearo, L.C.: Analysis of formal and informal networks and the role of internal actors in organizations: a comparative study. Res Soc Dev **10**(4), e35910412783–e35910412783 (2021)
10. Borgatti, S.P., Everett, M.G., Freeman, L.C.: Ucinet for Windows: Software for Social Network Analysis. Harvard, MA: Analytic Technologies (2022)
11. Bradley, B., Henry, S., Blake, B.: When can negativity mean success? Gender composition, negative relationships and team performance. Small Group Res **52**(4), 457–480 (2021)
12. Brandes, U., Wagner, D.: Analysis and visualization of social networks. In: Graph Drawing Software, pp. 321-340. Springer https://doi.org/10.1007/978-3-642-18638-7_15 (2004)
13. Bryan, L.L., Matson, E., Weiss, L.M.: Harnessing the power of informal employee networks. McKinsey Quarterly **4**, 44 (2007)
14. Byrne, J., Clarke, L., Van Der Meer, M.: Gender and ethnic minority exclusion from skilled occupations in construction: a Western European comparison. Constr. Manag. Econ. **23**(10), 1025–1034 (2005)
15. Casciaro, T., Lobo, M.S.: Competent jerks, lovable fools, and the formation of social networks. Harv. Bus. Rev. **83**(6), 92–99 (2005)
16. Combs, G.M.: The duality of race and gender for managerial African American women: Implications of informal social networks on career advancement. Hum. Resour. Dev. Rev. **2**(4), 385–405 (2003)
17. Cox, D. Informal Networks, Institutions and the 'Soap Opera' Constraint. Working Paper (1999)
18. Cross, R.: The Hidden Power of Social Networks: Understanding How Work Gets Really Done in an Organization. Harvard Business School Press (2004)
19. Cross, R., Prusak, L., Parker, A.: Where work happens: The care and feeding of informal networks in organizations. Institute for Knowledge-based Organizations, Cambridge, MA (2002)
20. Datcher, L.: The impact of informal networks on quit behavior. The review of Economics and Statistics, 491–495 (1983)
21. De Mooij, M., Hofstede, G.: Convergence and divergence in consumer behavior: implications for international retailing. J. Retail. **78**(1), 61–69 (2002)
22. De Toni, A.F., Nonino, F.: The key roles in the informal organization: a network perspective analysis. Learn. Organ. **10**(17), 86–103 (2010)
23. Donelly, M., Hillman, A., Stancliffe, R.J., Knox, M., Whitaker, L., Parmenter, T.R.: The role of informal networks in providing effective work opportunities for people with an intellectual disability. Work **36**(2), 227–237 (2010)
24. Durbin, S.: Creating knowledge through networks: A gender perspective. Gend. Work. Organ. **18**(1), 90–112 (2011)
25. Dwyer, S., Richard, O.C., Chadwick, K.: Gender diversity in management and firm performance: The influence of growth orientation and organizational culture. J. Bus. Res. **56**(12), 1009–1019 (2003)

26. Eboiyehi, C.O., Fayomi, I., Eboiyehi, F.A.: From exclusion to discrimination: Gender inequality in the senior management of Nigerian universities. Issues Edu Res **26**(2), 182–205 (2016)
27. Eliot, T.S.: Notes towards the Definition of Culture. Faber & Faber (2010)
28. Falkenberg, L., Herremans, I.: Ethical behaviours in organizations: Directed by the formal or informal systems? J. Bus. Ethics **14**(2), 133–143 (1995)
29. Farh, J.L., Hackett, R.D., Liang, J.: Individual-level cultural values as moderators of perceived organizational support–employee outcome relationships in China: Comparing the effects of power distance and traditionality. Acad. Manag. J. **50**(3), 715–729 (2007)
30. Faulkner, W., Kleif, T.: One size does not fit all! Gender in/exclusion in a rural community-based ICT initiative. J Adult Continuing Edu **11**(1), 43–61 (2005)
31. Gale, A., Cartwright, S.: Women in project management: entry into a male domain? A discussion on gender and organizational culture–part 1. Leadersh. Org. Dev. J. **16**(2), 3–8 (1995)
32. Gerecke, G.A., House, G.: Update: TMT national culture, demographic heterogeneity, and profit change using LTO-WVS. Res Bus Econ J **8**, 1 (2013)
33. Gómez-Borja, M. Á., Lorenzo-Romero, C.: Are the users of social networking sites homogeneous? A cross-cultural study. Frontiers in Psychology, 6 (2015)
34. Hallett, T.: Symbolic power and organizational culture. Soc Theory **21**(2), 128–149 (2003)
35. Hofstede, G.: The History of a Social Invention. In: Bleichrodt, N., Drenth, P.J.D. (eds.) Contemporary issues in cross-cultural psychology, pp. 4–20. Swets & Zeitlinger Publishers (1979)
36. Hofstede, G.: Empirical models of cultural differences (1991)
37. Hofstede, G., Minkov, M.: VSM 2013. Values survey module (2013)
38. Hofstede, G., Minkov, M.: VSM 2013. Values survey module manual (2013)
39. Horak, S., Afiouni, F., Bian, Y., Ledeneva, A., Muratbekova-Touron, M., Fey, C.F.: Informal networks: Dark sides, bright sides, and unexplored dimensions. Manag. Organ. Rev. **16**(3), 511–542 (2020)
40. Huang, C., Yi, H., Chen, T., Xu, X., Chen, S.: Networked environmental governance: formal and informal collaborative networks in local China. Policy Stud **43**(3), 403–421 (2022)
41. Janićijević, N.: The mutual impact of organizational culture and structure. Econ. Ann. **58**(198), 35–60 (2013)
42. Kakabadse, N.K., Figueira, C., Nicolopoulou, K., Hong Yang, J., Kakabadse, A.P., Özbilgin, M.F.: Gender diversity and board performance: Women's experiences and perspectives. Hum. Resour. Manage. **54**(2), 265–281 (2015)
43. Kegen, N.V.: Science networks in cutting-edge research institutions: Gender homophily and embeddedness in formal and informal networks. Procedia Soc. Behav. Sci. **79**, 62–81 (2013)
44. Khan, F., Sohail, A., Sufyan, M., Uddin, M., Basit, A.: The effect of workforce diversity on employee performance in Higher Education Sector. J Manage Info **6**(3), 1–8 (2019)
45. Khosla, N.: The ready-made garments industry in Bangladesh: A means to reducing gender-based social exclusion of women? J Int Women's Stud **11**(1), 289–303 (2009)
46. Kochan, T., et al.: The effects of diversity on business performance: Report of the diversity research network. Human Resource Management: Published in Cooperation with the School of Business Administration, The University of Michigan and in alliance with the Society of Human Resources Management **42**(1), 3–21 (2003)
47. Krackhardt, D., Stern, R.N.: Informal networks and organizational crises: An experimental simulation. Social psychology quarterly, 123–140 (1988)
48. Kuzu, Ö.H., Özilhan, D.: The effect of employee relationships and knowledge sharing on employees' performance: An empirical research on service industry. Procedia Soc. Behav. Sci. **109**, 1370–1374 (2014)

49. Leung, K., Bhagat, R.S., Buchan, N.R., Erez, M., Gibson, C.B.: Culture and international business: Recent advances and their implications for future research. J. Int. Bus. Stud. **36**(4), 357–378 (2005)
50. Lien-Tung, C., Cheng-Wu, C., Chen-Yuan, C.: Are educational background and gender moderator variables for leadership, satisfaction and organizational commitment? Afr. J. Bus. Manage. **4**(2), 248–261 (2010)
51. Maranto, C.L., Griffin, A.E.: The antecedents of a 'chilly climate' for women faculty in higher education. Human Relat **64**(2), 139–159 (2011)
52. McGuire, G.M.: Gender, race, and the shadow structure: A study of informal networks and inequality in a work organization. Gend. Soc. **16**(3), 303–322 (2002)
53. Mehng, S.A., Sung, S.H., Leslie, L.M.: Does diversity management matter in a traditionally homogeneous culture? Equality Divers Incl Int J **38**(7), 743–762 (2019)
54. Mehta, C.M., Strough, J.: Sex segregation in friendships and normative contexts across the life span. Dev. Rev. **29**(3), 201–220 (2009)
55. Milligan, C., Littlejohn, A., Margaryan, A.: Workplace learning in informal networks. In: Reusing open resources, pp. 115–125. Routledge (2014)
56. Mitchell, M.A., Yates, D.: How to use your organizational culture as a competitive tool. Nonprofit World **20**(2), 33–34 (2002)
57. Mun, E., Jung, J.: Change above the glass ceiling: Corporate social responsibility and gender diversity in Japanese firms. Adm. Sci. Q. **63**(2), 409–440 (2018)
58. Naqvi, S., Ishtiaq, M., Kanwal, N., Butt, M.U., Nawaz, S.: Impact of gender diversity on team performance: The moderating role of organizational culture in telecom sector of Pakistan. Asian J Soc Sci Humanit **2**(4), 228–235 (2013)
59. Naseer, S., Donia, M.B.L., Syed, F., Bashir, F.: Too much of a good thing: The interactive effects of cultural values and core job characteristics on hindrance stressors and employee performance outcomes. Hum. Resour. Manage. **59**(3), 271–289 (2020)
60. Rank, O.N.: Formal structures and informal networks: Structural analysis in organizations. Scand. J. Manag. **24**(2), 145–161 (2008)
61. Rohim, A., Budhiasa, I.G.S.: Organizational culture as moderator in the relationship between organizational reward on knowledge sharing and employee performance. J Manage Dev **38**(7), 538–560 (2019)
62. Rowlinson, S.: Matrix organizational structure, culture and commitment: a Hong Kong public sector case study of change. Constr. Manag. Econ. **19**(7), 669–673 (2001)
63. Sabharwal, N.S., Henderson, E.F., Joseph, R.S.: Hidden social exclusion in Indian academia: gender, caste and conference participation. Gend. Educ. **32**(1), 27–42 (2020)
64. Schneid, M., Isidor, R., Li, C., Kabst, R.: The influence of cultural context on the relationship between gender diversity and team performance: A meta-analysis. Int J Hum Res Manage **26**(6), 733–756 (2015)
65. Schneider, B., Smith, D.B.: Personality and organizational culture. In Personality and organizations Psychology Press, pp. 371–394 (2004)
66. Soda, G., Zaheer, A.: A network perspective on organizational architecture: performance effects of the interplay of formal and informal organization. Strateg. Manag. J. **33**(6), 751–771 (2012)
67. Stevenson, W.B.: Formal structure and networks of interaction within organizations. Soc. Sci. Res. **19**(2), 113–131 (1990)
68. Stevenson, W.B., Radin, R.F.: The minds of the board of directors: the effects of formal position and informal networks among board members on influence and decision making. J. Manage. Governance **19**(2), 421–460 (2014). https://doi.org/10.1007/s10997-014-9286-9
69. Taylor, D.G., Lewin, J.E., Strutton, D.: Friends, fans, and followers: do ads work on social networks?: how gender and age shape receptivity. J. Advert. Res. **51**(1), 258–275 (2011)

70. Van den Brink, M.: Scouting for talent: Appointment practices of women professors in academic medicine. Soc. Sci. Med. 72(12), 2033–2040 (2011)
71. Van den Brink, M., Benschop, Y.: Gender in academic networking: The role of gatekeepers in professorial recruitment. J. Manage. Stud. **51**(3), 460–492 (2014)
72. Van Der Heijden, B., Boon, J., Van der Klink, M., Meijs, E.: Employability enhancement through formal and informal learning: an empirical study among Dutch non-academic university staff members. Int. J. Train. Dev. **13**(1), 19–37 (2009)
73. Van Emmerik, I.H., Euwema, M.C., Geschiere, M., Schouten, M. F. Networking your way through the organization: Gender differences in the relationship between network participation and career satisfaction. Women Manage Rev (2006)

Care and Social Support Networks: An Exploration of Theoretical and Methodological Approaches

Francisca Ortiz Ruiz[(✉)] [iD]

Millennium Institute for Care Research (MICARE), Santiago, Chile
franortizruiz@gmail.com

Abstract. Care and social support networks have been crucial to understanding the connections among people in a society at any period. This research aims to explore those two concepts' theoretical and methodological notions, revising how they had been understood through different studies. Care and support networks are not consistently differentiated. They provide resources, information, help, and even tension between people in their daily lives. I propose a social interdependence approach toward care and social support networks. Specifically, I propose that care networks are the available and perceived ones, while support networks are the actual networks measured in an event or over a specific period.

Keywords: Care networks · Social support networks · Older people

1 Introduction

Care networks (CNs) and social support networks (SSNs) have been problematized in the literature from many different approaches. This paper draws on the information collected in 2022 in Santiago de Chile on older people's care and social support networks. The main objective is to understand broadly the composition of these networks are and how they change over time. As the main two concepts of this research are care and support networks, the first stage was to clarify each one accordingly to the literature. It is found that this distinction is rarely made, leading to confusion and a need in disentangling their differences. This study aims to advance in filling this gap.

In the social network analysis literature, distinctions are made between 'available', 'perceived' and 'actual' networks [1]. These three different aspects are related to three different questions: First, the network of helpers available to an ego can be determined, as well as the latent network of alters that can be inactive at any particular time. Second, questions can be focused on the networks of people perceived by an ego as eventual helpers. The third option accounts for a specific event or an episode and help received/provided on that specific occasion, giving the actual network. The first two options of available and perceived networks sometimes overlap, depending on the case. The available network is not always well-known by an ego, as sometimes there are

© The Author(s), under exclusive license to Springer Nature Switzerland AG 2023
A. Antonyuk and N. Basov (Eds.): NetGloW 2022, LNNS 663, pp. 191–209, 2023.
https://doi.org/10.1007/978-3-031-29408-2_13

helpers they do not know (e.g., support from the state with a salary bonus for health-related problems). The perceived, available and actual networks are different. Then, if one considers those differences, these networks can be measured precisely.

Care and social support networks are differentiated by the type of network: perceived or actual. Care networks are those available alters (of any nature) that can provide support, meaning they are the available and perceptive network. Social support networks are those actual alters (of any nature) that provide support in a specific event, meaning they conform to the actual network. This approach tries to understand society's social interdependence in a broader sense, incorporating both notions – care and social support networks. In this article, the manner in which this distinction could work in empirical research will also be presented.

In this study, I aim to identify different theoretical, methodological and empirical approaches to the notions of care and social support networks. To my knowledge, this differentiation has not always been specified by studies that use both notions. In this article, I revise the notions of care and social support networks and show the differences of the two concepts through empirical research.

This text is divided into six main sections. First, a panoramic from the literature is presented to show how care has been understood. Second, following the same logic, social support definitions are shown; each of those parts ends with a summary. Third, an exploratory proposal is described to differentiate both notions from theoretical and methodological approaches. Then, the context of this research is explained regarding some methodological decisions and fieldwork. In the fifth section, results are shown with comparisons between perceived and care networks. These are the first results of this research. Finally, the text is concluded with some reflections on limitations.

2 Care Networks

2.1 The Approaches to Care in the Literature

The care literature generally agrees that this concept is broader and more complex than others. As it has been said, 'the very concept of "care" overflows with paradoxes and ambivalence' [2: 31]. This is related to the concept being embedded in the ontological concept of society, as well as linked to many aspects that a researcher could focus on such as activities, relationships, emotional work, formality, informality, institutionalisation, among others. It is a polysemic and varied concept [2–4]. Three different but complementary approaches have been developed towards understanding care: (1) as a value to reach justice and equality in society, (2) as activities and relationships and (3) like an intricate network, or links and nodes.

As a Value to Reach Justice and Equality. The concept of care has been used for years to promote equality and justice in an ethics of care approach. In the thirties, Carol Gilligan conveyed that caregiving is not a natural characteristic of women; on the contrary, it is learned and is socially constructed as an intrinsic value of women [5, 6]. She argued that maternity and all the care work done for children and others were obscured for years, which left any discussions about care as belonging to the private sphere. Her conclusions exhibited a different proposal: to highlight the value of care as one that should be defended, diffused and required as universal. Care answers to universal circumstances, as it is something that every human needs at some stage in their lives, and it should be included in any democracy. This approach is used to understand that all individuals are relational and interdependent. Democracy without care is patriarchal and not equal.

Gilligan's approach was then used by Joan Tronto in the fifties [7]. She again emphasised the role of care in society, and that nobody could avoid the duty of care. The author extended the idea of care as a public value, where taking care of others is an ethical responsibility that any human being has at some point in their lives [8, 9]. It involves acknowledgement, respect, solidarity, fraternity and justice. Joan Tronto suggested that the ethics of care is democratic, as it is a distribution of responsibilities among the people. The state should not only provide the care but should also promote a public debate about assigning those responsibilities.

More recently, the discussion has included aspects not previously considered, such as self-care and caring for the environment. Victoria Camps [5] suggested that caring for oneself is the principle of any active life governed by the principle of moral rationality and is a sure way of facing the world, behaving and establishing relationships with others. In addition, taking care of the environment is an achievement that could help everyone in the world, as responsibility to one another and as part of the interrelationship between people and nature. It is a voice for continuity between human beings and nature that, in addition to expanding the horizons of ethics and politics, has followed the same principles from the beginning.

A care crisis has been recognised in more recent years; there are not enough people, institutions and organisations covering all the needs of society, especially for those in vulnerable groups. The Care Collective proposes that universal care is an ideal society where care is at the centre, all people are responsible and hands-on care work maintains communities and the world [2]. The relevance of this approach is that we need to take charge of it to answer a necessity that is currently becoming a crisis.

As Activities and Relationships. The concept of care crisis has led to social organisations that manage care at a macro level, look into daily activities and practices, like cleaning the dishes, going to the supermarket or giving money. Arriagada has defined this phenomenon as 'interrelationships between economic and social care policies. It is about the way of distributing, understanding and managing the need for care that sustains the functioning of the economic system and social policy' [10: 59]. This notion is also sustained by the idea of a dyad, in which one individual need another.

Caregiving is not only about helping, as it could create difficult and stressful scenarios for the caregiver or the person receiving the care. The Care Collective reflects that: '(…) The word care in English comes from the Old English caru, meaning care, concern, anxiety, sorrow, grief, trouble – its double meanings clearly on display (…) can be both challenging and exhausting.' [2: 32]. Evelyn Nakano Glenn highlighted a coercive aspect of caregiving linked to its obligation [11], which may negatively impact those involved.

On the one hand, some authors have defined care as primarily a summary of activities [12, 13] that are both directly and indirectly focused on the well-being of the receivers. The main purpose here is facilitating and maintaining daily life [10]. For Huenchuan [14], these actions are more specifically received by those who do not have personal autonomy and need others to take care of the essential acts of daily life. Others emphasised that these are not only activities but also relationships [6, 11]. Care focuses on satisfying the physical and emotional necessities of others [6]. Meanwhile, Glenn defined caregiving as maintaining the daily lives of others intergenerationally [11]. In addition, care also exceeds aspects such as 'geographical limits (geographical distance/proximity), spheres of action (public/private), relational (biology/selection), moral (personal interest/altruism), physical (dependence/autonomy) and temporal (life–time/working time), among others.' [15: 128].

As a Driver of Network Formation. In the social network analysis literature, the majority of studies focused on care is related to health settings, problematising medical care and how the networks were activated to deal with illness caregiving [16–19]. In addition, they focused on the care providers, informal caregivers and the burden associated with their actions to improve someone's life [16, 20, 21]. These studies had a crucial public impact, as they served to identify points of intervention for reducing stress levels for those receiving care. From this perspective, care has been studied as a relational notion, especially in health-related settings where one person has illness and needs. A couple of studies concerned childcare settings and networks of parents who were in position to help with their care [22, 23].

Other studies are more concentrated on the social support networks of people in situations of need, and care disappears from that discussion. Caregiving is taking care of someone who is ill and needs some help with activities in daily life [19, 23, 24]. For example, Bilecen and Cardone [24] defined three dimensions of care: 1) social support, 2) information exchange and financial resources, and 3) care relationships. The last one comprises relationships in different settings, which could include health care, care of older people, childcare, emergencies, moving home and inside the household.

Recently, Grabham studied how women balance caregiving responsibilities and work-defined care. In her own words, 'Care is notoriously difficult to define, involving a wide range of activities, relationships and emotional work inside and outside of the home that is oriented to another person's daily and/or developmental needs. (…) Care networks describe complex, collective and multi-institution arrangements involving different forms of informal and formal care provided by family, friends, schools, nurseries and care workers.' [25: 55]. Therefore, caregiving is not only a dyadic relationship; it also involves other people, institutions and organisations at different levels and interdependent nodes.

2.2 Summary of the Notion of Care

Care is helping others with their needs at any stage of their lives. Care is a value and a relational concept. If it is shared, it supports justice and equality in society and can involve people, institutions and organisations. In a society of social interdependence, care is at the centre and involves sharing responsibilities. Caregiving involves activities, emotional work, relationships and specific actions according to the care-receivers and the context. Taking care of others involves multidimensional and multi-institutional arrangements that can be understood as interdependent networks of links and nodes.

There is a diversity of care settings, some of which are immediately recognisable as educational, healthcare institutions and childcare; however, caregiving is broader than those settings as a part of the everyday life of any person.

3 Social Support Networks

3.1 The Approaches to Social Support in the Literature

Social support is one of the critical concepts of the social network theories. From its origin, social network analysis has measured social support, making many advances available in studies of this concept. Two different approaches to this concept will be briefly presented.

As a Way to Cope with Poverty. It has been agreed that social support is a multidimensional and sophisticated concept, and its definition varies significantly in the literature [26–29]. These definitions do not always consider all the dimensions available to define social support, and each author chooses to emphasise different aspects. Many studies have used social support as their primary concept [30], making it more relevant to specify which dimensions are included in their analyses. For example, Cobb defined social support as 'information leading the subject to believe that he is cared for and loved, esteemed and a member of a network of mutual obligations.' [31: 300]. Meanwhile, another author considered social support as 'any action or behaviour that functions to assist the focal person in meeting his personal goals or in dealing with the demands of any particular situation.' [32: 411]. In addition to the positive dimension of social support, studies have considered the negative dimension, such as tensions arising from the difficulties of supporting the poor. Despite this ambivalence, many authors seem to agree that social support networks are a mechanism to cope with difficulties [33].

This approach focuses on social support as help provided to those in need, which could vary by context. Wellman [34] collected several studies from varied settings, including France, Japan, China and Chile. In all of them, the idea of support was presented as help provided to others and as a way to cope with difficulties. For example, the social support networks of people in neighbourhoods with high poverty levels were studied [35]. His research highlighted that in that specific country, like others in Latin America, people lived in an economy of survival. The author explained that 'Resources circulate among people who are long-time friends and neighbours, and who often are kinfolk. They live close by, visit each other, and meet often.' [35: 175]. Accordingly, his results indicated the existence of strong ties between neighbours through an intense exchange of resources, which creates internal cohesion and helps people survive economic and political crises in the short term. However, Espinoza [35] noted that this environment does not produce better opportunities for social mobility, which in the long term means no improvement to the economic conditions. Therefore, this author understood these SSNs helped them to cope with and survive poverty.

Another study that showed this idea of social support as a way to survive is by Edin and Lein [36]. The study took place in Boston, Charleston, Chicago and San Antonio in the United States and used interviews with a random sample of single mothers living in those cities. Edin and Lein attempted to understand whether the poverty line at that moment was in line with reality. They concluded that it was not; however, the study provided evidence showing how these mothers only survived on their low incomes because of their SSNs. Therefore, we again see that SSNs are a means of coping with and surviving difficult experiences.

Aa a Source of Stress and Constraints. Until now, social support has been conceptualised as help and a way to cope with poverty. Nonetheless, some researchers have highlighted that social support could also be a source of stress and constraints to a person's daily life. This idea implies that SSNs can be problematic due to the person's context; they could be in a poverty situation regarding the resources they have to give to others.

González de la Rocha [37, 38] has reviewed studies from sociology and anthropology about life in poverty and households in different Latin American countries since the 1980s. She uncovered the dominance of the myth of survival that 'has been constructed together with a fairy-tale, which emphasises adaptation, solidarity and reciprocity as the major tools for surviving in conditions of poverty' [38: 46]. This idea has been maintained in different studies over time. Nevertheless, according to the author, it needs to be challenged, as it helps maintain an aggressive and violent economic blow to people's lives. She concluded that 'the resources-of-poverty model is no longer empirically or theoretically viable. Instead, a poverty of resources characterises the stage we are in today. The poverty of resources is the outcome of labour exclusion and the persistence and intensification of poverty' [37: 86]. González de la Rocha highlights macrostructures' effects on persistence and poverty intensification. Accordingly, those produce the poverty of resources for the person.

Offer [39] reviewed different empirical studies that used ethnography to study low-income families in the United States that had different racial and ethnic backgrounds. From her comparison, the data shows how the idea of reciprocity as an obligation in social relationships can produce two mechanisms in these families: exclusion and withdrawal. In her words, 'In the context of poverty, it can be a burden and source of relational stress that leads to the demise of social relationships. Furthermore, it is argued that poverty can make it difficult for individuals to maintain relations with others and participate in SSNs because they do not have many resources to share and reciprocate' [39: 788]. Therefore, this study established the reciprocity of SSNs as a burden – in other words, a constraint.

Menjívar's study [40] examined Salvadorian immigrants' networks in the United States. She stated that 'The assumption that extended networks represent a "survival strategy" to ameliorate the effects of poverty needs to be re-examined, perhaps even reversed, for the very conditions that informal networks supposedly mitigate may impede resource-strapped people from assisting one another' [40: 156].

A more recent study by Lubbers et al. [41] showed that the mobilisation of social support could create stress and be a constraint. They interviewed 61 people (40 women and 21 men) living below the Spanish poverty threshold in Barcelona. The dynamics of formal and informal support were measured. They found that 'while kin had important support functions of loyalty and care, chronic poverty gradually depleted these resources (…) and jeopardised relationships' [41]. Furthermore, 'The norm of kin obligation raised expectations that put family relationships under pressure' [41: 85]. Therefore, the SSNs were two different things simultaneously: help to an ego and a cause of stress and constraints.

3.2   Summary of the Notion of Social Support

Social support networks can be viewed from a broad perspective. According to Barrera and Ainlay, support networks are more social than psychological [30]. These are not static networks; they are dynamic and evolve as a result of a variety of experiences. Many factors influence them, including socioeconomics, geography, cultural environments, demographics, birth, family, age, context, generation and historical moments. Therefore, this perspective offers a social and relational approach to defining support networks.

A person's SSN can vary depending on various factors. It could change accordingly with the situation/context, and the individual characteristics of ego like age and gender. The author previously addressed this issue in other articles [42, 43].

The SSN is more than just a pool of resources or a safety net for dealing with problems; it is also a daily network of help, protection and sustenance that keeps society running on a micro level. On the other hand, social support can be the source of pressures and limits in a person's life, depending on the intersectionality of structural inequalities such as gender, socioeconomic status, race, ethnicity and age, among others. Then there are SSNs that are not only active whenever they are needed for coping – they are also constructed and anchored throughout people's lives.

4   A Proposal: The Social Interdependence

Social interdependence references the fact that at some point, all people need others in order to live. The level of dependency can be higher at certain stages of life, as for older

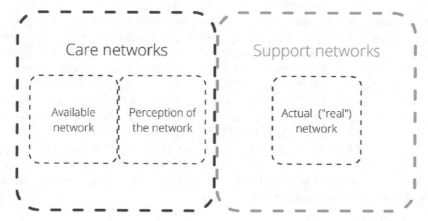

Fig. 1. Difference between care and support networks; linked to available, perceived and actual networks. Visual representation done by the author.

people with disabilities or chronic health problems. In analyses of existing research on support networks, a clarification is needed as to whether an available, perceived or actual (real) network has been measured. This distinction is crucial for this study because social interdependence is based on the existence of people's CNs and SSNs. The proposal here is to differentiate between care and social support so their measures are distinctive, as it is shown in Fig. 1.

In terms of definition, social support and care are similar notions. They provide resources, information, help, and even tension among people in their daily lives. Also, they both involve emotions and are embedded in specific situations. Furthermore, social support and care were (and will continue to be) shaped by the life histories of ego, alters and the contexts. The difference between those two concepts is their nature, which is not always identified.

Even though those two concepts are close together, there is one difference: care corresponds with an ontological idea, allowing us to understand reality's form and nature. It is more related to social justice and recognising a care crisis context, so it is the available and perceived help. Meanwhile, social support is a more practical idea, allowing us to see the reality of the relationship between the receiver and the giver. It is what has been exchanged/given/received in a specific period.

Care networks include the available and perceived networks that measure available alters (of any nature) that can provide support. SSNs are actual networks that measure actual alters (of any nature) that provide support in specific events. Even though this approach distinguishes between both concepts, it understands that they are sides of the same coin and are, therefore, complementary. It focuses on information about the same person's CNs and SSNs, which gives us access to a broader understanding of his or her life.

This form of differentiation between CNs and SSNs could be helpful for measuring them. As has been previously noted, both concepts are complex. They have different dimensions (i.e., material aid or emotional help) and levels (i.e., ego, alters, organisations); they are also embedded in context, social norms, life history and longitudinal dynamics.

Social support and care networks have many dimensions depending on the context. In this research, the operationalization of this concept included seven dimensions: 1) Economic: any help that involves money, material, o resources. 2) Emotional: support on an intimate and emotional level; people with whom to talk when there are problems and difficulties in the life of an ego. 3) Advice: people giving or receiving guidance and recommendations. 4) Physical assistance: practical help in domestic activities in daily life, like buying food, cleaning the house, or getting a bath. 5) Conversational: People within whom an ego talks informally about daily life without involving in deep conversations. For example, it could be talking with the neighbours about an event that will happen for Christmas or the current news on TV. 6) Positive interaction: People whom an ego would invite to enjoy the moment, like having some tea, watching movies together, having holidays or going out for dinner. 7) Negative interactions: People who create difficulties and tensions in the ego's life.

An example to explain the distinction between the available, perceived and actual networks is when we try to understand where people go for help with an activity related to their health (i.e., doctor's appointments, tests, medications and the like). That question could be asked in three different ways, with three different answers and results (see Fig. 2). First, it could concern the available network to help this ego; alters who are available less frequently and appear to be inactive at any particular time would be the latent network [44–46]. The broad question here is, 'Do you know if there are some institutions, or organizations or people that can help you with this activity?' This focus is on the broad sense of help and could refer to institutions or organisations at diverse levels. The question here is more about the ego's awareness of his or her network, which is different from the actual network [1, 47]. Second, the question could be focused on the networks perceived by an ego as eventual helpers, in which case we ask, 'If you need to ask for some help related to your health, who will you ask?' Finally, the third option is to ask for a specific event or episode and the help provided and received on that specific occasion. This approach has been broadly used around an episode lived by ego [48, 49]. Then, the question is distinctive: 'Would you tell me the last episode related to your health? … Now, could you tell me who helped you in that specific event? Anyone else?'.

The first two options of available and perceived networks sometimes overlap, depending on the situation. The available network is not always well-known by an ego, as sometimes there are helpers they do not know (like support from the state with a salary bonus for health-related problems). Therefore, whether the perceived network contains that information depends upon whether the ego has that knowledge and perceives it as a clear option.

Apart from the distinction between types of networks, there is a controversy related to the accuracy of respondents when reporting their ties – whether or not they are trustworthy, or perhaps they forgot something; perhaps the egos would like to impress the interviewer. We could problematise this instead as 'mistakes' with the data collection. They could also be appreciated for what they are – the interviewees' answers. Perhaps it is not the researchers' mission to know if people are telling the truth, because those are the interviewers' perceptions. They are part of the narratives counted by the egos at that moment, and as such, they are accurate [50].

Therefore, following the presented definitions of 'available', 'perceived' and 'actual' networks, and complemented with the notions of care and social support, we identified corresponding indicators and formulated a list of questions for measuring them. The questions are represented in Fig. 2 as a way to differentiate between the concepts of care networks and support networks.

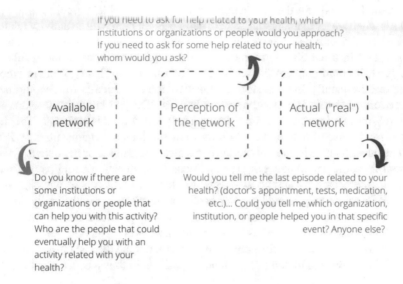

Fig. 2. Questions to collect information on available, perceived and actual networks. Visual representation done by the author with examples of questions.

5   Methodology

5.1   Context of This Research: Santiago De Chile, Latin America

This article is a part of broader research that seeks to understand older people's care and support networks in a community centre in Santiago de Chile. In the last decades in Latin America, there have been many changes in the structure and dynamics of families with a tendency to 'live longer'. One of the biggest problems in this context is that families are incapable of supplying the necessities of those individuals who need more care and support. As a consequence, the care that has been given by women decreases in each home. Then a more extensive network of care is required.

In recent years, Chile has become one of the countries with the highest replacement rate in Latin America, especially if we consider the southern cone [51, 52]. In this context, older people find themselves under a neoliberal pension regime that has increased long-term structural inequities [53, 54]. Additionally, previous studies highlight that, to this day, family members are the caregivers and fundamental supporters of older people [55]. In this context, the supportive care that older people receive and deliver becomes increasingly relevant.

5.2   Participants

The participants are all the older people at that centre and the professionals who work there, so those who did not answer the interview were still available to be nominated.

The participants are:

- Older people in the centre: all people over 60 years of age who are registered and who attend the centre on a regular basis. The list of registered older people is delivered to the responsible investigator by the centre's staff.
- Professionals: the people who work in the centre and whose profession is taking care of the ageing. There are seven people in this centre, including the director. Regarding their levels of training, they are all professionals in the health area: kinesiology, nursing and medicine. The centre director is a kinesiologist, and the other six people have the other professions. The director's role is coordinating all the centre's activities. The other professionals distribute the tasks of caring for older people and carrying out internal activities. In addition, they conduct training workshops and physical and mental therapy sessions.

The centre was chosen by searching the poorest districts in the Metropolitan Region that have the highest proportion of people over 60 years of age, according to the 2017 census. Older people who did not declare an interest in participating, and those who could not consent and had mental health-related issues, were excluded from the sample. In total, 10 people were excluded.

The sample of participants comprised 57 people from one community centre in Santiago de Chile, with seven professionals, 40 older women and 10 older men. During the first stage of data collection, in May 2022, 10 older people were not available to be interviewed and two refused to answer, meaning that 24% of the data was missing. A total of 38 surveys were answered by the older women (32) and men (6). In the sample, on average the amount of pension per month was $240,000 Chilean pesos, which is approximately 260 euros, with a minimum of 128 euros and a maximum of 537 euros. Only five of the sample had no chronic disease, and the age range of years was between 68 and 94.

5.3 Data Collection and Instrument

This research used a mixed method approach [56], with two instruments are used for data collection. The first is the name generator applied in the same way on three different occasions. It is a questionnaire to build the care and support networks of all the people in the day centre (older people and professionals) (see Table A.1 in the Appendix). This questionnaire with the same questions was applied at three points in time to investigate changes in the networks; the months are May, July and September 2022. The second instrument is a survey for people who agree to in-depth interviews which are planned for the end of 2022. Then, the fieldwork is designed to take place between May and December 2022. In this paper, only the data from the first wave collected during May of 2022 in Santiago de Chile is analysed. The interviews are expected to obtain information about the complete care and social support inside the institution.

The operationalisation of the concepts of care and support networks were done following the framework proposed in this paper. Then, the perceived and actual networks along seven specific dimensions were mapped using questions shown in Table A.1. In this way, the social interdependence framework in this empirical research was defined.

In addition, questions about some characteristics of ego and alters were included. The order of the questions was: 1) questions about characteristic of egos, 2) questions about each dimension of the perceived networks (or CNs), and finally, 3) questions about each dimension of the actual network (SSNs). As the complete network is measured, questions about ego characteristics were followed by question about alter characteristics. Then, the available network would be all the people in the Community centre.

For the analysis, the networks were treated separately. In other words, separate directed networks were created for each dimension, allowing us to compare them. The links are directed to those the older people nominated. In the network visualizations, professionals (seven nodes in total) are distinguished with another colour, and nodes of small size indicate people who were nominated, but did not have an opportunity to be interviewed. In this paper, we only compared the perceived and actual networks. The available network is not included to reduced complexity. It is to be consider in future research.

6   What Do We Find in the Data? Comparing Perceived and Actual Networks

As previously stated, the idea of social interdependence framed this empirical research. Figure 3 shows four dimensions of care networks and SSNs: economic, emotional, advice and physical assistance. Many professionals identify themselves with a higher socioeconomic class than the older people; then, we can identify how they are constantly being asked for money or resources, even though, in the actual networks these links almost completely disappear, meaning that the last time older people asked for money, they did not ask people inside the community centre. In the emotional dimension, we can see more similar patterns between perceived and actual networks: the professionals are nominated more often. Regarding the emotional and physical assistance, the perceived network again is centred around the professionals, although this changes in the actual networks. It could be hypothesized that a stronger relationship with trust needs to be built before asking for advice or physical assistance from others. Also, it may be the case that older people asked for assistance along these dimensions, or provided it, to family members or friend outside the centre.

The results show how different the perceived and actual networks are along each of these dimensions. We have many isolated nodes in the actual network, which some researchers could interpret as a symbol of isolation or loneliness. However, we must consider that they may have been people from outside the centre. In addition, we can identify that the professionals were the most nominated in regard to these four dimensions. Meanwhile, not many names of older people were reported, so the links are more from older people to professionals and not among themselves. Following these results, I could hypothesise that older people use the community centre as a place for interactions and receiving health-related help, rather than as a place to construct personal relationships.

The following Fig. 4 represents the three other dimensions of networks: conversational, positive and negative interactions. In the first two, the professionals lose their central role, and now we see more extensive main components with more links among

Perceived networks   Actual networks

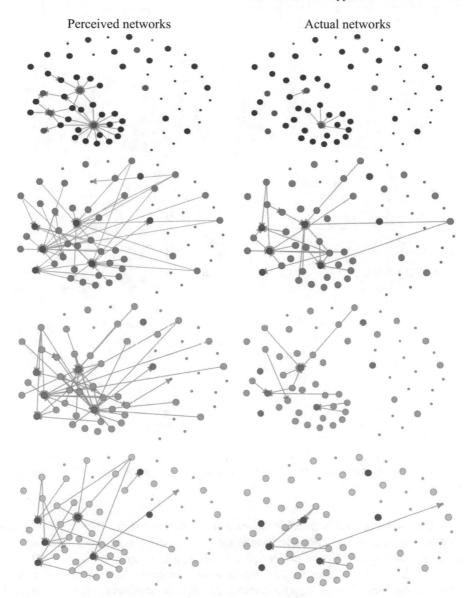

Fig. 3. Economic, Emotional, Advice and Physical Assistance dimensions of the perceived and actual networks reported by older people. The left side represents perceived networks; the right side represents the actual networks. Each dimension follows the order in the title.

older people. These two main components are reduced considerably in the actual networks, which could reinforce the idea that those networks are outside the centre. Negative interactions were nominated by only seven people, while the actual network is composed of only one link.

Perceived networks   Actual networks

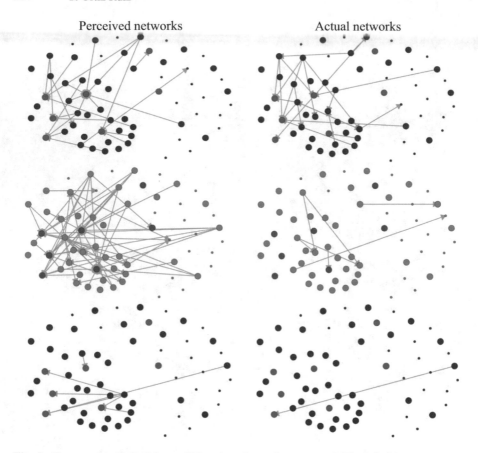

Fig. 4. Conversational, Positive and Negative, dimensions reported. The left side represents perceived networks; the right side represents the actual networks. Each dimension follows the order in the title.

A QAP test was used to examine the level of correlation and then similarity between the perceived and actual networks except for the negative interactions dimension as the corresponding actual network contained only one link, so it was not possible to apply a QAP test (see Table 1). The results show that the highest correlation was in the economic dimension. Next, a moderate correlation in emotional and conversational dimensions was reported. Finally, there was a low correlation for the dimensions of advice, physical assistance and positive interactions.

It is interesting to focus on how some dimensions differ more than others; advice, physical assistance and positive interactions result in more different networks. These three dimensions may require a stronger relationship between ego and alters to be nominated. The conversational and emotional dimensions had a moderate level of correlation. Meanwhile, the economic dimension was the one with the highest correlation between the perceived and actual networks. Accordingly, it can be hypothesized that older people

Table 1. Correlation between perceived and actual networks – QAP Test results Note: *p < 0.001. Replications: 10000.

		Actual network					
		Economic	Emotional	Advice	Physical assistance	Conversational	Positive
Perceived network	Economic	0.418					
	Emotional		0.344				
	Advice			0.229			
	Physical assistance				0.248		
	Conversational					0.355	
	Positive						0.178

tend to ask for resources or money people outside the community centre, e.g., family members or childhood friends.

7 Conclusion

In this paper, an analysis of how care and support networks have been understood as similar notions through different studies was presented. Care and support networks provide resources, information, help and even tension among people in their daily lives. I applied a social interdependence approach toward care and social support networks. Specifically, I proposed that care networks are the available and the perceived networks, while the support networks are the actual networks measured for specific events or specific periods. Considering social interdependence as a theoretical and methodological approach works from an empirical point of view, as has been shown in this paper. A theoretical clarification like the one proposed have methodological consequences, and ultimately influences the results and conclusions obtained in search of clarification.

The obtained results highlight the differences found between perceived and actual networks; or in other words, among CNs and SSNs. This difference can also be tracked through its specific dimensions. The most similar networks between each other were found at the economic dimension, followed by the emotional and conversational dimensions. Furthermore, the advice, physical assistance and positive interaction were found to be more different. Then, it can be hypothesized that these links exist in the actual network only when a more robust relationship is found. This aspect is something that needs further study.

For future studies, there is a need for a further discussion about which dimensions are considered for each concept. In this research, only seven dimensions were used; however, these dimensions depend on the research context. For example, in a city where the religious aspect of life is central, it could make sense to also ask for spiritual support. Whether to add it or not would also depend on the research objectives. Nonetheless, more research is needed that includes other dimensions that may be relevant for other contexts, for comparison. For example, in other countries it may be relevant to include a question about spiritual or professional advice.

This research has two limitations. First, it is not representative nationally, meaning the data collected here cover a group of older people, even though this allowed having deeper and denser data about this specific group. Second, this paper only presents the results obtained from the collection of the complete networks section, not about what happens with older people outside the community centre. It will be insightful to compare the presented results with the data of the CNs and SSNs outside the community centre.

Acknowledgements. This research was supported by the ANID Millennium Science Initiative Program (ICS2019_024). The research is titled 'Vivir cuidando: Usando métodos mixtos para comprender las redes de cuidados y soporte de personas mayores' ['Living taking care: using mixed methods to understand care and support networks of older people']. Without the support from the participants, this research would have not been possible. In addition, I would like to acknowledge the work done by assistant researchers who helped with the data collection: Trinidad Cereceda, Natacha Leroy, Ignacia Luco and Natalia Zipper. Finally, I would like to thank the reviewers and editors for their helpful comments and suggestions, which improved this article.

Appendix

Table A.1. Questions to measure care and social support networks. Note: All the questions here are phrased to ask for care/support links from alters to ego, although in the survey all of them were also phrased in the other way around.

Dimension	Perceived networks	Actual networks
Economic	If you have to ask for economic or material support (like money, food, clothes, rent, others), whom would you ask? Any others?	Think about the last time you needed economic or material support (like money, food, clothes, rent, and others). Whom did you ask? Any others?
Emotional	Whom would you ask if you need someone to hear about your worries and fundamental problems? Any others?	Think about the last time you needed someone to hear about your worries and fundamental problems. Whom would you ask? Any others?
Advice	If you need some advice on any topic of your life, whom would you ask? Any others?	Think about the last time you needed advice on any topic of your life. Whom would you ask? Any others?

(continued)

Table A.1. (*continued*)

Dimension	Perceived networks	Actual networks
Physical assistance	Whom would you ask if you need some help that demands physical assistance (like housework tasks, for example)? Any others?	Think about the last time you needed some help that demanded physical assistance (like housework tasks, for example). Whom would you ask? Any others?
Conversational	Whom would you ask if you want to talk with somebody about your daily life, news and other everyday conversations? Any others?	Think about the last time you needed to talk with somebody about your daily life, news and other everyday conversations. Whom would you ask? Any others?
Positive interaction	If you want to have a pleasant time with somebody (like going out for tea, celebrating an occasion, holidays or having a nice meal), whom would you invite? Any others?	Think about the last time you needed to have a pleasant time with somebody (like going out for tea, celebrating an occasion, holidays or having a nice meal), whom would you invite? Any others?
Negative interaction	Whom would you identify if you have to think about someone who may bring you some problems or tensions in your daily life? Any others?	Think about the last time that you needed someone who may bring you some problems or tensions in your daily life. Whom would you identify? Any others?

References

1. McCarty, C., Lubbers, M., Vacca, R., Molina, J.L.: Conducting personal network research. A practical guide. The Guilford Press, New York (2019)
2. The Care Collective. The Care Manifest. The Politics of Interdependence. London, United Kingdom, Verso (2020)
3. Duffy, M.: Making Care Count. A Century of Gender, Race and Paid Care Work. New Jersey: Rutgers University Press (2011)
4. Acosta, E.: Cuidados En Crisis. Universidad de Deusto, Mujeres Migrantes Hacia España y Chile. España (2015)
5. Camps, Victoria. Tiempo de Cuidados. Otra Forma de Estar en el Mundo. Barcelona, España, Arpa & Alfil Editores edn (2021)
6. Gilligan, C.: In a Different Voice: Psychological Theory and Women's Development. Harvard University Press, United States of America (1936)
7. Tronto, J.: Who Cares? How to Shape a Democratic Politics. Cornell University Press, London (1952)
8. Tronto, J.: Moral Boundaries. Routledge, New York (1993)
9. Tronto, J.: Caring Democracy. Markets, Equality and Justice. New York: New York University Press (2013)
10. Arriagada, I.L.: Crisis de cuidado en Chile. Revista de Ciencias Sociales Uso del tiempo, cuidados y bienestar en Desafíos de Uruguay y la región, XXII I(27), 58–67 (2010)
11. Glenn, E.N.: Forced to Care. Harvard University Press, United States of America (2010)

12. González, H.; Género, cuidados y vejez: Mujeres «en el medio» del trabajo remunerado y del trabajo de cuidado en Santiago de Chile. Revista Prisma Social **21**, 194–218 (2018)
13. Del Valle, T. La articulación del género y el parentesco desde la antropología feminista. In: V. Fons, A. Piella, & M. Valdés (Eds.), Procreación, crianza y género. Aproximaciones antropológicas a la parentalidad Barcelona, PPU, pp. 218–395). (2010)
14. Huenchuan, S.: Perspectivas Globales Sobre la Protección de Los Derechos Humanos de Las Personas Mayores, 2007–2013. Santiago de Chile: CEPAL (2014)
15. Gonzálvez, H.Y., Acosta, E.: Cruzar las fronteras desde los cuidados: La migración transnacional más allá de las dicotomías analíticas. In: Las fronteras del transnacionalismo. Límites y desbordes de la experiencia migrante en el centro y norte de Chile Santiago. Ocho libros, pp. 126–149. (2015)
16. De la Haye, K, Whitted, C, Koehly, L.M.: Formative Evaluation of the Families SHARE Disease Risk Tool among Low-Income African Americans. Public Health Genomics, 1–11 (2021)
17. Pomare, C., Long, J., Churruca, K., Ellis, L., Braithwaite, J.: Social network research in health care settings: Design and data collection. Social Networks **69**, 14–21 (2022)
18. Brewer, B., Carley, K., Benham-Hutchins, M., Effken, J., Reminga, J.: Exploring the stability of communication network metrics in a dynamic nursing context. Social networks **61**, 11–19 (2020)
19. Cornell, B.: Good health and the bridging of structural holes. Social networks **31**, 92–103 (2009)
20. Koehly, L.M., Marcum, C.S.: Multi-relational measurement for latent construct networks. Psychol. Methods **23**(1), 42–57 (2018)
21. Ashida, S., Marcum, C.S., Koehly, L.M.: Unmet expectations in Alzheimer's family caregiving: interactional characteristics associated with perceived under-contribution. Gerontologist **58**(2), e46–e55 (2018)
22. Bojarczuk, S., Mühlau, P.: Mobilising social network support for childcare: The case of Polish migrant mothers in Dublin. Social networks **53**, 101–110 (2018)
23. Adams, A., Madhavan, S., Simon, D.: Measuring social networks cross-culturally. Social networks **28**, 363–376 (2006)
24. Bilecen, B., Cardone, A.: Do transnational brokers always win? A multilevel analysis of social support. Soc Netw **53**, 90–100 (2018)
25. Grabham, E.: Women, Precarious Work and Care. Bristol Policy Press, Bristol (2021)
26. Barrera, M.: A method for the assessment of social support networks in community survey research. Connections **3**, 8–13 (1998)
27. Agneessens, F., Waege, H., Lievens, J.: Diversity in social support by role relations: a typology. Soc. Netw. **28**, 427–441 (2006)
28. Thompson, E., Futterman, A., Gallagher-Thompson, D., Rose, J., Lovett, S.: Social support and caregiving burden in family caregivers of frail elders. J. Gerontol. **48**(5), 5245–5254 (1993)
29. Wellman, B., Wortley, S.: Different strokes from different folks: community ties and social support. Am. J. Sociol. **96**, 558–588 (1990)
30. Barrera, M., Ainlay, S.: The structure of social support: A conceptual and empirical analysis. J. Community Psychol. **11**(2), 133–143 (1983)
31. Cobb, S.: Social support as a moderator of life stress. Psychosom. Med. **38**(5), 300–314 (1976)
32. Tolsdorf, C.: Social networks, support, and coping: an exploratory study. Fam. Process **15**(4), 407–417 (1976)
33. Finch, J., Okun, M., Barrera, M., Zautra, A., Reich, J.: Positive and negative social ties among older adults: measurement models and the prediction of psychological distress and well-being. Am. J. Community Psychol. **17**(5), 585–605 (1989)

34. Wellman, B.: Networks in the Global Village: Life in Contemporary Communities. Westview Press, Boulder (1999)
35. Espinoza, V.: Social network among the urban poor: inequality and integration in a Latin American city. In: Networks in the global village: life in contemporary communities, Westview Press, United States, pp. 147–184. (1999)
36. Edin, K., Lein, L.: Making Ends Meet: How Single Mothers Survive Welfare and Low-Wage Work. Russell Sage Foundation, New York (1997)
37. González de la Rocha, M.: From the resources of poverty to the poverty of resources? Latin American Perspectives, 28(4), 72–100 (2001)
38. González de la Rocha, M.: The construction of the myth of survival. Development and change, 38(1), 45–66 (2007)
39. Offer, S.: The burden of reciprocity: Processes of exclusion and withdrawal from personal networks among low-income families. Curr. Sociol. 60(6), 788–805 (2012)
40. Menjívar, C.: Fragmented Ties. Salvadoran Immigrant Networks in America. University of California Press (2000)
41. Lubbers, M., Valenzuela García, H., Escribano Castaño, P., Molina, J.L., Grau Rebollo, J.: Relationships stretched thin: social support mobilization in poverty. ANNALS AAPSS 689, 65–88 (2020)
42. Ortiz, F., Bellotti, E.: The impact of life trajectories on retirement: socioeconomic differences in social support networks. Social Inclusion 9(4), 327–338 (2021)
43. Ortiz, F.: Como usar QCA y SNA desde los métodos mixtos: una ilustración usando las redes de soporte en jubilados. Revista Redes. Revista Hispana del Análisis de Redes Sociales, 32(2), 201–211 (2020)
44. Perry, B., Pescosolido, B., Borgatti, S.: Egocentric Network Analysis: Foundations, Methods and Models. Cambridge University Press, Cambridge (2018)
45. Bernard, H., Johnsen, E., Killworth, P., Robins, S.: Estimating the size of an average personal network and of an event subpopulation: some empirical results. Soc. Sci. Res. 20(2), 109–121 (1979)
46. Morgan, D.; Neal, M.; Carder, P.: The stability of core and peripherical networks over time. Social Networks 19(I), 9–25
47. Cohen, S., Janicki-Deverts, G.L.: Can we improve our physical health by altering our social networks? Perspective on Psychological Science 4(4), 375–378
48. Hollstein, B., Wagemann, C.: Fuzzy-set analysis of network data as mixed method: personal networks and the transition from school to work. In: Mixed methods social networks research. Design and applications, pp. 237–268. Cambridge University Press, Cambridge (2014)
49. Small, M.: Someone to Talk, 1st Edn. Oxford University Press, New York (2017)
50. Riesman, C.: Narrative Methods for the Human Sciences. SAGE, California (2008)
51. ECLAC. Anuario Estadístico de América Latina y el Caribe. Santiago, Chile, Naciones Unidas: Comisión Económica para América Latina y el Caribe (2015)
52. Pérez, V., Sierra, F.: Biología del envejecimiento. Rev. Med. Chil. 137, 296–302 (2009)
53. Wisensale, J.: Global Aging and Intergenerational Equity. J Int Relat 1(1), 29–47 (2003)
54. Sojo, A.E.: Sistema Contributivo de Pensiones Como Locus de Rivalidad y de un Nuevo Pacto Social en Chile. CEPAL, Chile (2014)
55. Ortiz, F., Gonzálvez, H.: Cómo explicar la organización social de los cuidados en Chile: una aproximación al proceso de envejecer. In Vera, Antonieta (ed.) Malestar social y desigualdad en Chile. Santiago de Chile: Editorial Universidad Alberto Hurtado, (pp. 125–150) (2017)
56. Ortiz, F.: Métodos mixtos para el análisis de redes sociales. REDES 34(1), 74–86 (2023)

Author Index

A. Antonyuk and N. Basov (Eds.): NetGloW 2022, LNNS 663, p. 211, 2023.
https://doi.org/10.1007/978-3-031-29408-2

Printed in the United States
by Baker & Taylor Publisher Services